Svava Kim Wetzel

Folding and Unfolding Mechanism of Designed Ankyrin Repeat Proteins

Svava Kim Wetzel

Folding and Unfolding Mechanism of Designed Ankyrin Repeat Proteins

Südwestdeutscher Verlag für Hochschulschriften

Impressum/Imprint (nur für Deutschland/ only for Germany)
Bibliografische Information der Deutschen Nationalbibliothek: Die Deutsche Nationalbibliothek
verzeichnet diese Publikation in der Deutschen Nationalbibliografie; detaillierte bibliografische
Daten sind im Internet über http://dnb.d-nb.de abrufbar.
 Alle in diesem Buch genannten Marken und Produktnamen unterliegen warenzeichen-, marken-
oder patentrechtlichem Schutz bzw. sind Warenzeichen oder eingetragene Warenzeichen der
jeweiligen Inhaber. Die Wiedergabe von Marken, Produktnamen, Gebrauchsnamen,
Handelsnamen, Warenbezeichnungen u.s.w. in diesem Werk berechtigt auch ohne besondere
Kennzeichnung nicht zu der Annahme, dass solche Namen im Sinne der Warenzeichen- und
Markenschutzgesetzgebung als frei zu betrachten wären und daher von jedermann benutzt
werden dürften.

Verlag: Südwestdeutscher Verlag für Hochschulschriften GmbH & Co. KG
Dudweiler Landstr. 99, 66123 Saarbrücken, Deutschland
Telefon +49 681 37 20 271-1, Telefax +49 681 37 20 271-0
Email: info@svh-verlag.de
Zugl.: Zürich, University of Zürich, Dissertation, 2008

Herstellung in Deutschland:
Schaltungsdienst Lange o.H.G., Berlin
Books on Demand GmbH, Norderstedt
Reha GmbH, Saarbrücken
Amazon Distribution GmbH, Leipzig
ISBN: 978-3-8381-0506-2

Imprint (only for USA, GB)
Bibliographic information published by the Deutsche Nationalbibliothek: The Deutsche
Nationalbibliothek lists this publication in the Deutsche Nationalbibliografie; detailed
bibliographic data are available in the Internet at http://dnb.d-nb.de.
 Any brand names and product names mentioned in this book are subject to trademark, brand
or patent protection and are trademarks or registered trademarks of their respective holders.
The use of brand names, product names, common names, trade names, product descriptions
etc. even without a particular marking in this works is in no way to be construed to mean that
such names may be regarded as unrestricted in respect of trademark and brand protection
legislation and could thus be used by anyone.

Publisher: Südwestdeutscher Verlag für Hochschulschriften GmbH & Co. KG
Dudweiler Landstr. 99, 66123 Saarbrücken, Germany
Phone +49 681 37 20 271-1, Fax +49 681 37 20 271-0
Email: info@svh-verlag.de

Printed in the U.S.A.
Printed in the U.K. by (see last page)
ISBN: 978-3-8381-0506-2

Copyright © 2011 by the author and Südwestdeutscher Verlag für Hochschulschriften GmbH &
Co. KG and licensors
All rights reserved. Saarbrücken 2011

Die vorliegende Arbeit wurde von der Mathematisch-naturwissenschaftlichen Fakultät der Universität Zürich im Frühlingssemester 2008 als Dissertation angenommen.

Promotionskomitee:
Prof. Dr. Andreas Plückthun
PD Dr. Ilian Jelesarov
Prof. Dr. Ben Schuler

Svava K. Wetzel

Department of Biochemistry
University of Zürich
Winterthurerstr. 19
CH-8057 Zürich
Switzerland

Zum Andenken an meinen Vater, für Jascha & Enrico

Abstract

The 'protein folding problem', i.e. understanding the detailed steps of the protein folding pathways is one of the most challenging topics in biochemistry. Combining theoretical with experimental research became indispensable for the investigation of the mechanisms in protein folding reactions.

Here, the biophysical characterization of designed ankyrin repeat (AR) proteins and the analysis of the experimental data by theoretical models are presented. Repeat proteins comprise repeating structural units that form linear multirepeat arrays. The non-globular elongated structure distinguishes them fundamentally from most studied proteins and makes repeat proteins interesting for folding studies.

In previous work, a full-consensus ankyrin repeat had been designed. This idealized ankyrin repeat served as a model repeat to study the folding of ankyrin repeat proteins.

In the first part of the thesis, a series of six designed ankyrin repeat proteins (DARPins with 1 up to 6 full-consensus repeats and two terminal capping repeats) was constructed and analyzed using light scattering, circular dichroism and fluorescence spectroscopy. The longer DARPins (> 5 repeats) were so stable that they could not be fully unfolded by GdnHCl. Therefore, kinetic analysis was performed only with the shorter three DARPins. The chevron plots clearly show deviations from the two-state model of folding. Cooperative folding models suggest a three-state mechanism with on-pathway intermediate; however, the full analysis of the equilibrium data of all six DARPins as well as the kinetic data of the smaller three DARPins using an Ising-like model provides further insights into possible parallel folding pathways. This model allows drawing an energy landscape for the DARPins at each denaturant concentration. Importantly, at high denaturant concentrations, an intermediate state with folded repeats and just the capping repeats unfolded would be almost as stable as the fully unfolded protein.

In the second part of the thesis a series of DARPins without capping repeats was constructed as well as variants with modified C-terminal capping repeats. The importance of the capping repeats for solubility and for avoiding aggregation was demonstrated. The C-terminal capping repeat was observed to denature first in MD simulations. Thus, a C-cap conferring similar solubility and even higher stability was designed, predicted *in silico* and validated experimentally. NI_3C showed a stable intermediate in experimental equilibrium unfolding. On the basis of the simulation results, this intermediate was interpreted to represent a conformation with four folded repeats and the unfolded C-cap. To validate this hypothesis, equilibrium unfolding experiments of NI_3 without C-cap and a NI_3C variant with a more stable C-cap (Mutant 5 and 6) were performed. The results confirmed the interpretation that the wild-type C-cap is less stable and therefore prone to unfold first. An engineered C-cap (Mutant 5 and 6) further improved the stability of the DARPin variants and did not show this equilibrium intermediate. This C-cap can now be used for biotechnological applications.

Zusammenfassung

Das 'Problem der Proteinfaltung', d.h. das Verstehen der einzelnen Schritte des Proteinfaltungsweges, ist eine der anspruchvollsten Fragestellungen der Biochemie. Daher ist die Kombination von Theorie und Praxis ist unerlässlich, um die Mechanismen von Proteinfaltungsreaktionen zu erforschen.

In dieser Dissertation wird die biophysikalische Charakterisierung von künstlich entwickelten Ankyrin Repeat Proteinen (Designed Ankyrin Repeat Proteins, DARPins) und deren Analyse durch theoretische Modelle vorgestellt. Repeat Proteine bestehen aus sich wiederholenden Struktureinheiten in einer nicht-globulären, gestreckten Anordnung, was sie grundlegend von den meisten in der Proteinfaltung untersuchten Proteinen unterscheidet und sie zu einer interessanten Proteinklasse für Faltungsstudien macht.

Die hier verwendeten Ankyrin Repeats basieren auf einer Konsensussequenz von natürlichen Ankyrinproteinen und dienen damit als Modell für die Faltung dieser Proteine.

Im ersten Teil der Dissertation wurden sechs verschiedene DARPins mit einem bis sechs internen Repeats sowie jeweils zwei terminalen Capping Repeats konstruiert und mittels Lichtstreuung, Zirkulardichroismus- und Fluoreszenz-Spektroskopie analysiert. Die längeren DARPins (> fünf repeats) waren jedoch so stabil, dass es nicht möglich war, sie in 8 M GdnHCl komplett zu entfalten; aus diesem Grund wurden die kinetische Experimente lediglich mit den kürzeren drei DARPins durchgeführt. Die Chevron Plot Analysen zeigen deutliche Abweichungen vom Zwei-Zustands-Faltungsmodell, weshalb von den kooperativen Faltungsmodellen ein Drei-Zustands-Mechanismus mit on-pathway Intermediat suggeriert wird. Jedoch liefert die globale Analyse der Gleichgewichtsdaten aller sechs DARPins zusammen mit den kinetischen Daten der kleineren drei DARPins anhand eines Ising-ähnlichen Modells weiterführende Einblicke in mögliche parallel ablaufende Faltungswege. Dieses Modell erlaubt die Erstellung einer Energielandschaft zu jeder Denaturierungsmittel-Konzentration für die DARPins. Gemäss der Energielandschaft wäre bei hoher Denaturierungsmittel-Konzentration ein Intermediärzustand mit gefalteten internen Repeats und entfalteten Capping Repeats fast so stabil wie das komplett entfaltete Protein.

Im zweiten Teil dieser Dissertation wurden sowohl DARPins mit modifizierten C-terminalen capping repeats als auch ohne capping repeats konstruiert. Die Bedeutung der capping repeats für die Löslichkeit und die Aggregationstendenz wurde bewiesen. Da sich der C-terminale Capping Repeat in Molecular Dynamics Simulationen zuerst entfaltete, wurde ein C-cap entworfen, welcher dem Protein ähnliche Löslichkeit und sogar höhere Stabilität verlieh. Die C-cap-Stabilität wurde *in silico* ermittelt und durch Experimente bestätigt. NI_3C zeigte ein stabiles Intermediat in experimentellen Gleichgewichtsübergängen. Anhand der Simulationsergebnisse wurde angenommen, dass dieses Intermediat eine Konformation darstellt, in der vier Repeats gefaltetet sind, während der C-cap entfaltet ist. Zur Validierung dieser Hypothese wurden Gleichgewichtsübergänge mit NI_3 ohne C-cap und einem NI_3C Varianten mit einem stabileren C-

Zusammenfassung

cap (Mutante 5 und 6) durchgeführt. Die Ergebnisse bestätigten die Interpretation, dass der Wildtyp C-cap weniger stabil ist und daher dazu tendiert, zuerst zu entfalten. Ein engineered C-cap (Mutante 5 und 6) konnte die Stabilität der DARPins weiter erhöhen und zeigte kein Equilibrium-Intermediat.

This thesis is based on the following documents, which are cited later on in the text.

List of Documents

I Wetzel, S. K., Settanni, G., Kenig M., Binz H. K. & Plückthun, A. (2008). Folding and Unfolding Mechanism of Highly Stable Full-Consensus Ankyrin Repeat Proteins. *J. Mol. Biol.* **376**, 241-257.

II *Interlandi, G., *Wetzel, S. K., Settanni, G., Plückthun, A. & Caflisch, A. (2008). Characterization and further stabilization of designed ankyrin repeat proteins by combining molecular dynamics simulations and experiments. *J. Mol. Biol.* **375**, 837-854.

* contributed equally

III Merz, T., Wetzel, S. K., Firbank, S., Plückthun, A., Grütter, M. & Mittl, P. R. E. (2008). Stabilizing ionic interactions in a full consensus ankyrin repeat protein. *J. Mol. Biol.* **376**, 232-240.

IV Li, L. W., Wetzel, S., Plückthun, A. & Fernandez, J. M. (2006). Stepwise unfolding of ankyrin repeats in a single protein revealed by atomic force microscopy. *Biophys. J.* **90**, L30-L32.

Table of contents

1. Introduction   1
1.1 *Repeat proteins*   1
1.1.1 Ankyrin repeat proteins and DARPins   2
1.2 *Principles of Protein Folding*   3
1.2.1 Two-state model   5
1.2.1.1 Thermodynamics and kinetics in the two-state case   5
1.2.2 Three-state models   7
1.2.2.1 Kinetic three-state models   8
1.2.3 Modular Ising Model   8
1.3 *Protein Folding studies of natural AR proteins*   9
1.4 *Protein Folding studies of designed consensus repeat proteins*   11
1.4.1 Designed AR, LRR and Armadillo repeat proteins   11
1.4.2 Designed TPR proteins   12
1.5 *The aim of the project*   13

2. Results   14
2.1 *Folding and Unfolding Mechanism of Highly Stable Full Consensus Ankyrin Repeat Proteins*   14
2.2 *Characterization and Further Stabilization of Designed Ankyrin Repeat Proteins by Combining Molecular Dynamics Simulations and Experiments*   32
2.3 *Stabilizing Ionic Interactions in a Sulfate Binding Ankyrin Repeat Protein*   51
2.4 *Stepwise unfolding of ankyrin repeats in a single protein revealed by atomic force microscopy*   61

3. Protocols   65
3.1 *unfolding and refolding curve*   65
3.2 *fast kinetic un- and refolding using a stopped-flow instrument*   66

4. Conclusions and Discussion   71
5. Outlook and Perspectives   73
6. References   77
7. Appendix   80
7.1 *Abbreviations*   80
7.2 *List of Plasmids*   81
7.3 *Equations used for fitting of cooperative folding models*   82
7.3.1 Equilibrium two-state fit   82
7.3.1.1 Chemical unfolding, fitting parameters m, Dm   82

7.3.1.2 Chemical unfolding, fitting parameters m, ΔG^0 82
7.3.1.3 Thermal unfolding, fitting parameters m, ΔH^0 83
7.3.2 Equilibrium three-state fit   83
7.3.2.1 Chemical unfolding, fitting parameters m_1, ΔG_1^0, m_2, ΔG_2^0 83
7.3.2.2 Chemical unfolding, fitting parameters m_1, K_1^0, m_2, K_2^0 84
7.3.3 Kinetics: Two-state chevron plot   84
7.3.4 Kinetics: Three-state chevron plot   85
7.3.4.1 on-pathway intermediate   85
7.3.4.2 off-pathway intermediate   85
7.3.4.3 triangular pathway intermediate   86

Introduction
1.1 Repeat proteins

Repetitive sequence segments occur in at least 14 % of all proteins (ranging from short amino acid repetitions to large repetitions containing multiple domains).[1] In the last two decades, a set of protein structures emerged that contain repeating structural units. These folds are termed solenoids, spirals, coils, supercoils, coiled folding domains and protein repeats, among other names.[2]

They can be found in all phyla, more commonly in eukaryotic organisms than in prokaryotic ones, and in metazoans more than in the rest of the eukaryotes. Apart from their high frequency among known sequences, repeat proteins have also many different functions for similar repeat types. The most common function is that of binding to proteins involved in processes like protein transport, protein-complex assembly and protein regulation.

A repeat protein comprises repeating structural units that form three-dimensional multirepeat assemblies. These assemblies are arranged in such a way, that the polypeptide chain forms either a linear array or a continuous superhelix, where the repeats are arranged about a common axis (see Figure 1).[2,3] Most repeat proteins exhibit therefore an elongated and nonglobular shape.

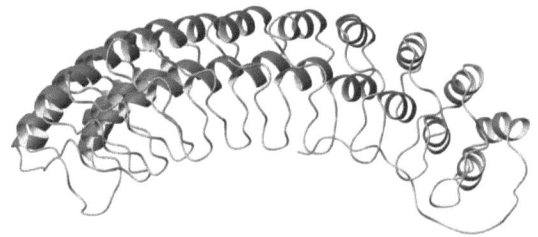

Fig. 1:
Structure of the D34 region of human Ankyrin-R (PDB code 1N11). This 12 AR region is one of the biggest AR proteins and illustrates the continuous superhelix.[4]

For some repeat types, there is no theoretical limit on the repeat number, as adding repeats is not sterically obviated. Such an "open" structure presents an expanded solvent-accessible surface that is ideal for binding large substrates as proteins and nucleic acids.

A repeat unit is formed of two, three or four elements of secondary structure linked by a short hairpin and the repeat families can be classified into three major structural types (Figure 2): all α (armadillo/HEAT and TPR-like repeats), mixed α/β (leucin-rich and ankyrin repeats) or all-β (β-propeller and β-trefoil with radial axis, β-strand repeats arranged along linear axis). The most prominent members are the ankyrin repeat (AR), armadillo repeat (ARM), leucine-rich repeat (LRR) and tetratricopeptide repeat (TPR).

Solenoid proteins exhibit the least complicated relationship between a sequence and the corresponding three-dimensional structure, i.e. the lack of contacts between units distant from each

Introduction

other in primary structure. The structures are stabilized by hydrophobic interactions between neighboring repeats (stacking interactions). The simple and fundamentally different structure from globular proteins makes the repeat proteins an interesting object for protein folding studies.

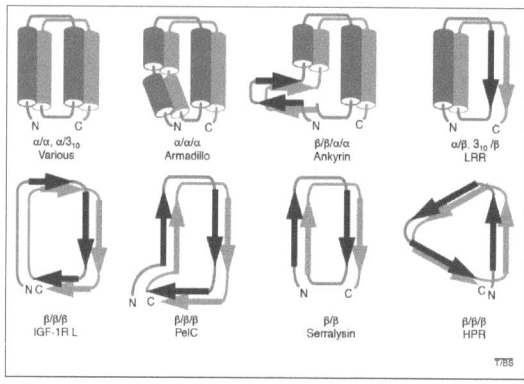

Fig. 2:

Schematic diagrams of repetetive structural units in solenoid proteins. Helices are shown as red cylinders, β-strands are shown as magenta arrows and the connecting loops are shown in green according to Kajava (2000). The next repeat is shown in grey.[2]

1.1.1 Ankyrin repeat proteins and DARPins

The ankyrin repeat (AR) is one of the most frequently observed protein sequence motifs. These repeats got their name from one of the proteins in which they were first found, the human erythrocyte protein ankyrin, containing 24 AR. Erythrocyte ankyrin attaches the spectrin skeleton (membrane skeleton) to band 3, the anion-exchange protein.[5]
AR proteins appear in nearly all phyla; apart from animals and yeast, they were also found in a plant protein and some prokaryotes and have even been noted in viruses such as poxviruses, variola, vaccinia and mouse mammary tumor virus.[6] The AR was found in > 3000 proteins, including cyclin-dependent kinase (CDK) inhibitors, transcriptional regulators, cytoskeletal organizers, spider toxins, in mitochondrial enzymes, nuclear cell cycle regulators, i.e. in various intra- and extracellular milieus (see also Fig. 3).
The ankyrin repeat consists of two antiparallel α-helices preceeded by a short turn and followed by a loop. They are arranged to a superhelical structure with an extended binding groove that is formed by the loops (anti-parallel β-sheet) and the helices.
The ubiquitousness of ankyrin repeat proteins and their high binding affinities led to the decision in our group to use the AR as protein scaffold for the generation of designed protein-binding molecules. The design of artificial ankyrin repeat proteins (DARPins) will be discussed in the subsection 1.2.5.1.

Introduction

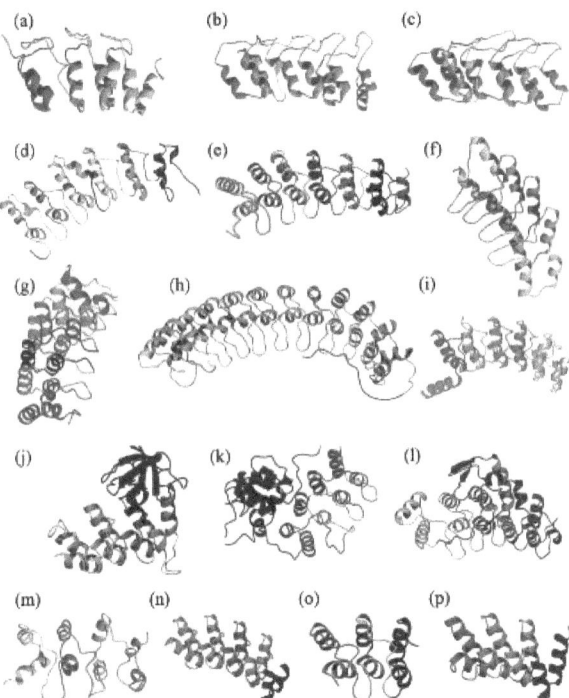

Fig.3:
Ankyrin repeat proteins with high-resolution structures in the PDB. The repeats have been colored differently to illustrate the packing interactions present in this fold. Natural AR proteins (a) – (m) and designed AR proteins (n) – (p) according to Mosavi et al.[7]

1.2 Principles of Protein Folding

Understanding the process how each newly synthesized peptide chain finds its way to a unique active conformation has fascinated scientists for decades. A solution of this "folding problem" is of enormous intellectual importance and would provide the missing link in the flow of information between a gene sequence and the 3-D structure of a protein, and then ultimately, its function.
Apart from the basic interest in the mechanism, several far-reaching implications in investigating protein folding do exist; for example, improving structure prediction algorithms and methods for *de novo* design of protein folds that did not evolve by nature, as well as fields ranging from medicine, as i.e. misfolded proteins of brain diseases, to nanotechnology, i.e. single molecule protein measurements.[8]

Introduction

The graph in Fig. 4 shows the number of publications in the field of protein folding per year, and pictures therefore the development in the field of understanding protein folding mechanisms over the last 45 years. While in the seventies there were very few protein folding studies available, in the beginning of the nineties the amount of publications in the field of protein folding increased tremendously.

In the late 50's, Christian B. Anfinsen denatured Ribonuclease A in 8 M urea with β-mercaptoethanol and discovered that, when renaturing the protein, it obtained back its biologically active state. This observation proved that the information necessary to specify the 3-D-structure of Ribonuclease A, is encoded in the amino acid sequence.[9,10] Cyrus Levinthal then addressed the question about how proteins can find their way to the native state. As there are an astronomical number of possible conformations for a protein to adopt, e.g. if we have only three conformations per amino acid, then we have $3^{100} \approx 10^{48}$ possibilities, the unbiased search through all of these possibilities would take by far too long for a protein to fold.[11] This phenomenon was called

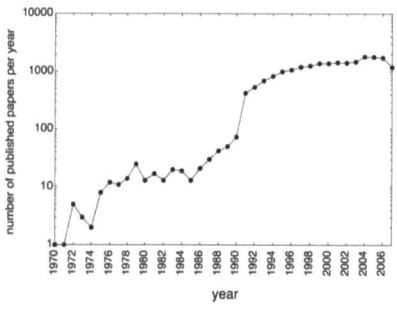

Fig. 4:

The number of manuscripts published per year (1970 – 2007) with the keywords 'protein folding' in either the title or the abstract. The data were taken from the Web of Science database.

the "Levinthal paradox". It is a logical step to argue that there must be defined pathways to facilitate the choices to reach the folded state.

After the awareness that proteins need to find the "right" pathway in order to reach their native structure, several models for the mechanism of folding have been proposed as summarized in Fig. 5. Anfinsen's original experiments demonstrated that proteins fold spontaneously and reversibly into their native conformation.

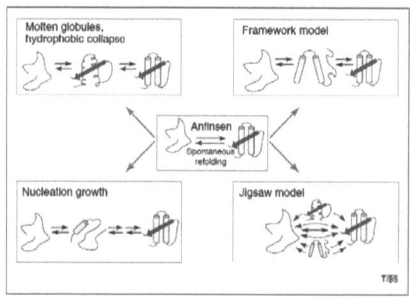

Fig.5: Early models for mechanisms of folding according to Radford (2000).[8]

Introduction

The nucleation growth model proposed that certain residues adjacent in sequence form a nucleus from which the native structure then is form in a sequential manner. The framework model, however, suggested that first all secondary structure elements form and that these then fold into the final native structure. The molten globule or hydrophobic collapse model assumes, that the proteins form first a collapsed intermediate species, and search then the native state. Finally, the Jigsaw model suggests that each protein molecule could fold by a different path.[8]

Later on with the development of new experimental techniques and theoretical methods, major advances were made in elucidating the folding mechanisms that are termed the "new view of folding".

Maybe the most important point in the perception of folding today is that there is not a single, specific folding pathway. Instead, a multidimensional energy landscape better describes the folding process. Such an energy landscape allows the protein to follow different ways to the native state, however some of them must be much more populated than others (Fig. 6). Such an energy landscape does not contradict the older classical models. However, the classical models concentrate on the definition of the intermediates, while the energy landscape aims to summarize the process in a more global view.

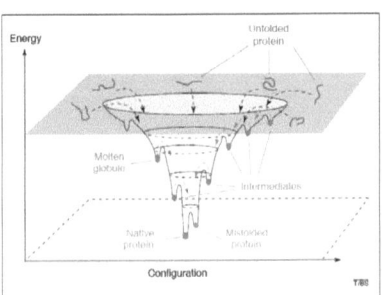

Fig. 6: Schematic diagram of a folding energy landscape. Denatured molecules at the top of the funnel might fold to the native state by a myriad of different routes, some of which involve transient intermediates (local energy minima) whereas others involve significant kinetic traps (misfolded states). For proteins that fold without populating intermediates, the surface of the funnel would be smooth.[8,12]

1.2.1 Two-state model

In the two-state model of protein folding, the proteins are believed to adopt only two states: native (N) and unfolded (U). Such a behavior has been observed for many small single-domain proteins with less than 100 residues.

1.2.1.1 Thermodynamics and kinetics in the two-state case

This scheme (1) describes the conformational equilibrium in the two-state model.

Introduction

$$U \underset{k_u}{\overset{k_f}{\rightleftharpoons}} N \qquad (1)$$

The reversible folding and unfolding is assumed to be a fully cooperative reaction such that no intermediates accumulate. The most important thermodynamic parameter characterizing a reversible equilibrium between the native and unfolded state is $\Delta G°$, the difference in the free energy between the folded and the unfolded state. $\Delta G°$ is related to the equilibrium constant $K°$ (by equations (2) and (3)).

$$\Delta G° = - RT \ln K° \qquad (2)$$
$$K° = [U] / [N] \qquad (3)$$

Plotting the spectroscopic signal of a certain protein as a function of the denaturant concentration yields the equilibrium transition curve (Fig. 7).

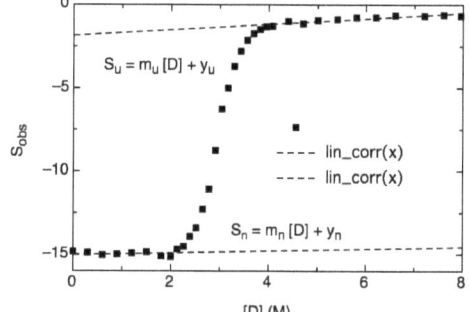

Fig. 7: Equilibrium unfolding transition of a two-state protein. The pre- and post-transition baselines are shown by dotted lines. The transition midpoint Dm is at 2 M urea. Adapted from Jelesarov [13]

$\Delta G°$ can be calculated by fitting the data points of the equilibrium transition using equation (4).

$$S_{obs}(D) = (S_U + m_U[D])f_U + (S_N + m_N[D])f_N \qquad (4)$$

k_u and k_f are the microscopic rate constants for the folding and unfolding reactions (see scheme (1)). The single observable macroscopic rate constant λ is derived as

$$\lambda = k_u + k_f \qquad (5)$$

and the equilibrium constant $K°$ is defined as

$$K° = k_f / k_u \qquad (6)$$

Introduction

Plotting the macroscopic rate constants k_{obs} as a function of the denaturant concentration, provides the so-called chevron plot, as shown in Fig. 8. The linear parts of the chevron plot are called folding and unfolding limbs. Linearly extrapolating the limbs to 0 M denaturant concentration (see equations (5) and (7)) yields the microscopic rate constants k_u and k_f in absence of denaturant, that are used to characterize the folding and unfolding kinetics of a certain protein. Generally, taking into account also more complex kinetics in which one has n states, for all k_{ij} between state i and state j (with $i=1,...,n$, $j=1,...,n$ and $j \neq i$), the relation

$$\ln k_{ij} = \ln k_{ij}^0 + m_{ij}[D] \qquad (7)$$

is assumed, i.e. a linear dependence of the logarithm of the rate on the denaturant concentration, where k_{ij} represents the denaturant-dependent rate constant, k_{ij}^0 the rate constant in the absence of denaturant. In the two-state case we have two states, the unfolded state 1 and the folded state 2. The unfolding rate k_u can be expressed as k_{12} and the folding rate k_f as k_{21}.

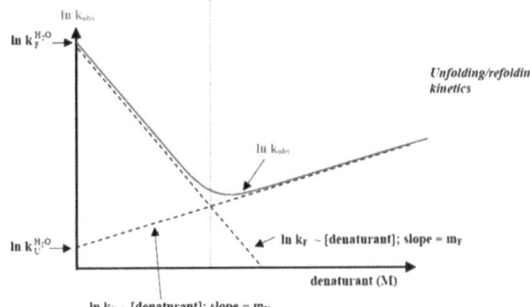

Fig. 8:
Chevron plot of a two-state protein. The extrapolation to 0 M denaturant is shown in dashed lines. Adapted from Jelesarov [13]

1.2.2 Three-state models

For some small proteins, as i.e. the four-helix bacterial immunity protein Im7[14,15] or the α-amylase inhibitor tendamistat[16], but commonly for larger proteins (more than 100 residues) three-state transitions involving an intermediate state have been detected (scheme (8)).

$$U \rightleftharpoons I \rightleftharpoons N \qquad (8)$$

The corresponding equations to derive $\Delta G°$ and further thermodynamic parameters are shown in the results.

Introduction

1.2.2.1 Kinetic three-state models

The possible three-state mechanisms are the triangular (9), the linear on-pathway intermediate (10) and the linear off-pathway intermediate mechanism (11).

$$U \underset{k_{NU}}{\overset{k_{UN}}{\rightleftharpoons}} N \quad \text{with} \quad U \underset{k_{IU}}{\overset{k_{UI}}{\rightleftharpoons}} I \underset{k_{NI}}{\overset{k_{IN}}{\rightleftharpoons}} N \tag{9}$$

$$U \underset{k_{IU}}{\overset{k_{UI}}{\rightleftharpoons}} I \underset{k_{NI}}{\overset{k_{IN}}{\rightleftharpoons}} N \tag{10}$$

$$I \underset{k_{UI}}{\overset{k_{IU}}{\rightleftharpoons}} U \underset{k_{NU}}{\overset{k_{UN}}{\rightleftharpoons}} N \tag{11}$$

Since all three mechanisms shown in scheme (9) to (11) give rise to experimental observable rate constants, it is impossible to exclude the triangular mechanism on the basis of spectroscopic measurements of the folding kinetics. However, special double mixing techniques as the interrupted refolding assay by Schmid[17] allow to distinguish between the on-path way and the off-pathway mechanism. The characteristic equations for the three-state mechanisms, i.e. linear quadratic equations involving the macroscopic and microscopic rate constants, and the analytical solutions of these equations are summarized in Buchner *et al*[18] and the derivation of the kinetic parameters are shown in the Results in chapter 2.1.

1.2.3 Modular Ising Model

The Ising model is an alternative to the classical cooperative models and gives more insights into possible folding mechanisms. Originally, this simple model is used to describe the effects of magnetization in the context of statistical mechanics. Ising-like models fully describe the

Introduction

thermodynamics of ferro-magnetic materials. Applied to the folding problem of DARPins an Ising-like model was developed and makes the following assumptions.
On the one hand, each repeat of a DARPin is considered as an independent two-state folding unit with a free energy of unfolding that depends linearly on the denaturant concentration $[D]$. On the other hand, adjacent folded repeats interact by a stabilizing potential J, whose magnitude is independent of $[D]$, but requires that both repeats are folded.
As our DARPins have capping repeats (symbolized by a rectangle) of different stabilities than the one of the internal full-consensus repeats (symbolized by an oval), two different free energy terms ΔG^0 and $\Delta G^{0'}$ have been introduced. A cartoon of a potential intermediate state of a DARPin with three internal repeats is shown in Figure 9. A detailed description of this model is found in the Results section 2.1.

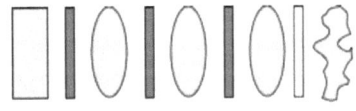

Fig. 9: Cartoon representation of an intermediate state of a DARPin with three internal full-consensus repeats. The folded capping repeat is shown as a rectangle, the folded internal repeats as ovals, an unfolded capping repeat is represented by a deformed rectangle, the interaction potential is shown as a bar and the missing interaction potential as an empty bar. The sum of the individual energy terms, describes the total energy E of this intermediate state: $E = \Delta G' + J + \Delta G + J + \Delta G + J + \Delta G + 0 + 0$.

1.3 Protein Folding studies of natural AR proteins

Several natural AR proteins have been previously investigated in folding studies.
One of the smallest AR proteins with a known structure is the tumor suppressor protein p16 with four AR. This protein is a member of the INK4 family of inhibitors of the cyclin D-dependent kinases, CDK4 and CDK6, that are involved in the key growth control pathway of the eukaryotic cell cycle. In tumors, p16 is frequently mutated and the cells can thus proliferate. An extensive thermodynamic[19] and kinetic analysis as well as a phi-value analysis allowed to state a sequential unfolding mechanism, where the N-terminal repeats 1 and 2 unfold first, and then the C-terminal repeats 3 and 4. The phi-value analysis is a method that has been introduced by Fersht, and makes use of thermodynamic and kinetic parameters in order to probe the conformation of the folding transition state of proteins.[20]
Unfortunately, refolding measurements were not possible, therefore no data exist about the folding mechanism.[21] The protein is very unstable and sensitive to point mutations with respect to its aggregation behavior. This vulnerability might provide one explanation for the frequency of p16 point mutations when analyzed in tumor-infected tissue.[19] A MD simulation study was consistent

Introduction

with the sequential unfolding mechanism and suggested two on-pathway intermediate states.[22] It was also found that the autonomous folding unit of the p16 protein consists of two repeats, i.e. the C-terminal repeats 3 and 4, p16C can fold independently, without the rest of the protein p16.[23]

The tumor suppressor protein p19 has five AR and belongs to the same family as p16. While urea-induced unfolding transitions by far-UV CD and phenylalanine fluorescence suggested a two-state mechanism, the unfolding of p19 followed by 2D ^{1}H-^{15}N HSQC spectra revealed a third species at moderate concentrations of urea.[24] Further intensive analysis of the folding and unfolding kinetics showed that this intermediate must be on-pathway and NMR H/D exchange experiments allowed structural speculations about this intermediate, i.e. the C-terminal repeats ANK3-5 are folded, while the N-terminal repeats 1-2 are unstructured.[25]

The third most intensively investigated natural AR protein for protein folding studies is the *Drosophila melanogaster* Notch receptor domain, a protein domain containing seven AR. The Notch protein signals by converting a transcriptional repression complex into an activation complex. The activity is modulated by interaction between the intracellular portion of the Notch receptor and a number of cytosolic and nuclear effector proteins, such as Suppressor of Hairless, Deltex, EMB-5 and Skip.[26] Numerous publications[27,28,29,30,31,32,33,34] give insight into many different aspects of this protein domain.

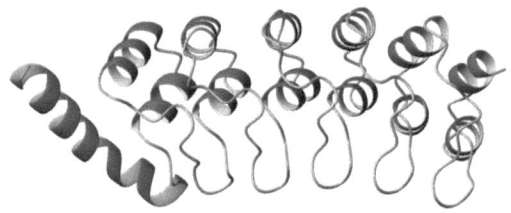

Fig. 10: The structure of the *Drosophila melanogaster* Notch receptor domain. The first repeat, which does not adopt an ankyrin-repeat fold, is on the left.

Sequence alignments show that only repeat 1 to 6 have high similarity, while the seventh repeat exhibits lower similarity to the consensus sequence. The crystal structure (Fig. 10) revealed that the seventh, C-terminal repeat adopts a regular ankyrin fold, but the first N-terminal repeat appears to be largely disordered.[26]

In another study with the mouse Notch homolog the higher stability of the C-terminal repeats was confirmed, as only repeats 4 to 7 were resistant to degradation during crystallization. The crystal structure showed a preservation of the typical ankyrin fold also for the poorly conserved seventh AR,[35] as seen in the *Drosophila* Notch domain.

The thermodynamic study of a series of deletion constructs containing subsets of the seven ankyrin repeats of the *Drosophila* Notch receptor showed that the stability increases linearly with the repeat number. Using an 1D Ising model for folding, an energy landscape for protein folding

Introduction

was determined.[36] To a good approximation, stabilities of each construct can be described as a sum of energy terms associated with each repeat. The magnitude of each energy term indicates that each repeat is intrinsically unstable but strongly stabilized by interactions with its nearest neighbors. A linear regression analysis of the stabilities versus repeat number yields an average stability of -2 kcal/mol for each single repeat and also shows that a single repeat of the *Drosophila* Notch receptor should be intrinsically unstable (+5.5 kcal/mol).[36]

Kinetic analysis of the full length construct of Notch, Nank1-7Δ (a 10 residue shorter construct with an N-terminal His$_6$ tag, where two prolines in the 7th repeat are omitted), provided insight into the folding and unfolding mechanism. Both the refolding and unfolding kinetics were described by two exponential phases, i.e. when plotting the rates against denaturant concentration, a non-linear v-shaped chevron plot is defined for both phases. These two chevron plots, together with the unfolding amplitudes (the extent of the signal change in a kinetic measurement) are consistent with a sequential three-state model. Thus, similar to the study of p16 and p19, the Notch receptor domain shows a complex folding mechanism, i.e. an intermediate state could be detected and is believed to occur on the path between the folded and the unfolded state (on-pathway).[37]

1.4 Protein Folding studies of designed consensus repeat proteins

1.4.1 Designed AR, LRR and Armadillo repeat proteins

In our group, three different repeats were used for consensus design, the ankyrin repeat[38], the leucin-rich repeat[39] and the armadillo repeat[40]. It was found when using AR for consensus design, highly stable binding molecules can be obtained that can then be used together with selection technologies such as phage display and ribosome display. Additionally, the AR proteins (DARPins) could be expressed in very high yields in soluble form in the cytoplasm (> 200 mg/l shake flask culture). The AR fold does not rely on cysteines or disulfide bridges, that require an oxidizing environment to be formed correctly. If AR proteins had to form disulfide bridges, expression would have to be directed to the periplasm, as is done when expressing antibody fragments.

Ankyrin repeat protein libraries were constructed, where the N- and C-terminal capping repeats from mouse GA-binding protein (GABPβ1 subunit) were used. The internal repeats were designed using a consensus strategy.[38,41,42] The design strategy was based on sequence alignments using SMART, GenBank and PFAM databases. The consensus sequence was defined by the most frequent residues and further refined using structural data. In this way potential target interaction residues and framework residues were defined. While the framework positions were left constant,

Introduction

the chosen interaction residue positions, which are part of the β-turn and the first α-helix, were allowed to vary. When I started my project, also a full-consensus AR just had been designed, where all the seven variable interaction residues had been fixed according the same design strategy. These design choices are discussed in chapter 2.1.

1.4.2 Designed TPR proteins

Similarly to our design of artificial ankyrin repeats, Regan and coworkers designed an artificial tetratricopeptide repeat (TPR).[43,44,45] Each repeat is composed of two helices and has almost the same number of residues as the ankyrin repeat. The designed TPR protein contains no loops and one additional "solvating" helix at the C-terminus that is important for solubility as the C-terminal capping repeat in the AR protein design (Fig. 11).[43]

Seven TPR proteins were constructed based on the same consensus containing one to ten consensus repeats (CTPR1 to CTPRa10).[46,47] Thermodynamic as well as kinetic measurements showed similar results to our full-consensus DARPin series. The TPR protein series was described by a one-dimensional Ising model.[47] This theoretical model was able to predict the protein stability in detail. The conclusion was that folding and unfolding of TPRs, and likely of all repeat proteins, does not conform to the all-or-nothing, folded-or-unfolded, two-state transition that is generally assumed for small globular proteins. Instead, the Ising description assumes the existence of partially folded configurations with significant statistical weight.[47]

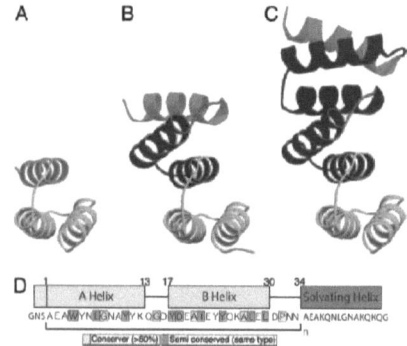

Fig. 11: Structure of the consensus tetratricopeptide repeat (TPR) proteins. (A) CTPR1. (B) CTPR2. (C) CTPR3. Repeat 1 is shown in yellow, repeat 2 in red, repeat 3 is blue and the solvating helix is green. (D) The amino acid sequences of CTPR proteins. Adapted from Main et al.[46]

This 1D Ising model inspired us to use a similar model for studying our DARPins series and these results will be shown in the Results section 2.1.

Introduction

1.5 The aim of the project

Using the full-consensus designed AR, I constructed a series of six proteins with one to six internal repeats, flanked by a C-terminal and N-terminal capping repeat according to the DARPin library construction. These fully designed AR proteins constitute an average structure for all natural AR and therefore serve as a model AR to extensively study the folding mechanism of AR proteins. Since natural ankyrin repeat proteins vary in sequence and each repeat has its intrinsic different stability value, no general conclusions for repeat proteins can be drawn. Our designed proteins, however, have identical internal repeats and represent thus a model system for the study of repeat proteins.

While we were interested to see the correlation between stability as well as folding rate and repeat number, we also wanted to analyze the folding mechanism in detail. The most interesting, but also most difficult question was to find informations about the structure of the intermediate states in the full-consensus DARPin folding mechanism.

Two collaborations completed this folding study. One the one hand, molecular dynamics unfolding simulations of three full-consensus DARPins were carried out and gave us suggestions for further experiments. MD simulations showed that the C-terminal capping repeat (C-cap) unfolded first. On the basis of this result, experiments were designed in order to study the unfolding of the C-cap. On the other hand, an alternative folding model (Ising-like) was developed to study the thermodynamics of the whole series of the six DAPRins as one data set.

Results I

Results

2.1 Folding and Unfolding Mechanism of Highly Stable Full Consensus Ankyrin Repeat Proteins (*I*)

This chapter describes the thermodynamic and kinetic measurements of the six full-consensus DAPRins NI_1C to NI_6C. The classical two- and three-state models were used to fit the data, but also an Ising-like model was chosen to describe the thermodynamic data set. Surprisingly, this model is also able to predict the kinetics of the proteins and can therefore suggest several possible folding pathways.

Wetzel, S. K., Settanni, G., Kenig M., Binz H. K. & Plückthun, A. (2008). Folding and Unfolding Mechanism of Highly Stable Full-Consensus Ankyrin Repeat Proteins. *J. Mol. Biol.* **376**, 241-257 (for the coloured version of the article refer to the publisher weblink http://dx.doi.org/10.1016/j.jmb.2007.11.046)

I

Folding and Unfolding Mechanism of Highly Stable Full-Consensus Ankyrin Repeat Proteins

Svava K. Wetzel[1], Giovanni Settanni[2], Manca Kenig[1], H. Kaspar Binz[1] and Andreas Plückthun[1]*

[1]Department of Biochemistry, University of Zürich, Winterthurerstrasse 190, CH-8057 Zürich, Switzerland

[2]MRC Centre for Protein Engineering, Hills Road, Cambridge CB2 0QH, UK

Received 3 July 2007; received in revised form 3 October 2007; accepted 16 November 2007
Available online 22 November 2007

Full-consensus designed ankyrin repeat proteins were designed with one to six identical repeats flanked by capping repeats. These proteins express well in *Escherichia coli* as soluble monomers. Compared to our previously described designed ankyrin repeat protein library, randomized positions have now been fixed according to sequence statistics and structural considerations. Their stability increases with length and is even higher than that of library members, and those with more than three internal repeats are resistant to denaturation by boiling or guanidine hydrochloride. Full denaturation requires their heating in 5 M guanidine hydrochloride. The folding and unfolding kinetics of the proteins with up to three internal repeats were analyzed, as the other proteins could not be denatured. Folding is monophasic, with a rate that is nearly identical for all proteins (~400–800 s^{-1}), indicating that essentially the same transition state must be crossed, possibly the folding of a single repeat. In contrast, the unfolding rate decreases by a factor of about 10^4 with increasing repeat number, directly reflecting thermodynamic stability in these extraordinarily slow denaturation rates. The number of unfolding phases also increases with repeat number. We analyzed the folding thermodynamics and kinetics both by classical two-state and three-state cooperative models and by an Ising-like model, where repeats are considered as two-state folding units that can be stabilized by interacting with their folded nearest neighbors. This Ising model globally describes both equilibrium and kinetic data very well and allows for a detailed explanation of the ankyrin repeat protein folding mechanism.

© 2007 Elsevier Ltd. All rights reserved.

Edited by F. Schmid

Keywords: protein folding; Ising model; ankyrin repeat proteins

Introduction

Repeat protein architecture does not rely on interactions between residues that are distant in sequence (long-range interactions), but stabilizing and structure-determining interactions are formed within a repeat and between neighboring repeats. This special feature, the modular nature of repeat proteins, makes them fundamentally different from globular proteins and, thus, interesting for testing experimental and theoretical views that have emerged from the study of globular proteins. Moreover, since repeat proteins are the only class of proteins that can be extended in size while still constituting a contiguous domain, unique questions about how folding and stability change with the number of repeats can be asked.

Repeat proteins constitute, next to immunoglobulins, the most abundant natural protein classes specialized in binding.[1,2] Ankyrin repeat (AR) proteins consist of repeating structural units (repeats) that stack together to form elongated nonglobular repeat domains. The AR is one of the most common

*Corresponding author. E-mail address: plueckthun@bioc.uzh.ch.
Present addresses: M. Kenig, Novartis Lek Pharmaceuticals, Kolodvorska 27, S1-1234 Menges, Slovenia; H.K. Binz, Molecular Partners AG, Grabenstrasse 11a, CH-8952 Zürich-Schlieren, Switzerland.
Abbreviations used: AR, ankyrin repeat; DARPin, designed ankyrin repeat protein; GdnHCl, guanidine hydrochloride; TPR, tetratricopeptide repeat; RCO, relative contact order.

0022-2836/$ - see front matter © 2007 Elsevier Ltd. All rights reserved.

protein sequence motifs. This 33-residue motif consists of a β-turn, followed by two antiparallel α-helices and a loop that reaches the turn of the next repeat[3] (see Fig. 1a).

Stability and kinetic folding studies of mostly natural AR proteins have been performed. The tumor-suppressor protein p16[6,7] unfolds in a sequential manner; first, both N-terminal repeats unfold, followed by the two C-terminal repeats. Molecular dynamics simulations have been carried out to study this in more detail.[8] The tumor-suppressor protein p19 shows an equilibrium intermediate, as well as three folding phases.[9] In a more detailed kinetic study, an on-pathway intermediate was detected, as well as a suggestion for its structure was made using NMR hydrogen/deuterium exchange.[10] Similarly as observed for the p16 protein, both N-terminal repeats 1 and 2 unfolded first, while repeats 3–5 were still folded. When dissecting the Notch receptor ankyrin domain from *Drosophila melanogaster*,[11,12] constructs from four to seven repeats were made, in which multiple repeats were deleted from either end or both ends, resulting in the finding that stability increased with repeat number. The longest construct has been used for kinetic folding studies,[13,14] and it was found that refolding and unfolding kinetics are best described by a sum of two exponential phases.

Equilibrium and kinetic studies were also conducted with myotrophin, a small four-repeat AR protein.[15,16] While the kinetics of the Notch ankyrin domain could be fitted by a sequential three-state model, myotrophin kinetics were assigned to a two-state model. Further analysis with single and double mutants showed that myotrophin follows parallel pathways, where folding is initiated either by the C-terminal repeat or by the N-terminal repeat.[17]

All these studies showed that the folding of AR proteins is not simply a cooperative process, but intermediate states do occur. However, they have all been carried out with natural proteins containing repeats of different sequences and stabilities. Hence, many results only describe the particular protein under study, and they can only partially and qualitatively test the effects of protein length on kinetics and thermodynamics. In addition, they give no indication on the intrinsic properties of the consensus AR.

We therefore intended to examine the folding and unfolding of designed ankyrin repeat proteins (DARPins) with identical repeats as a function of repeat number. The consensus AR represents an "average structure" of all of the natural ARs and will eliminate properties that only come about with particular sequences of individual repeats. The DAR-

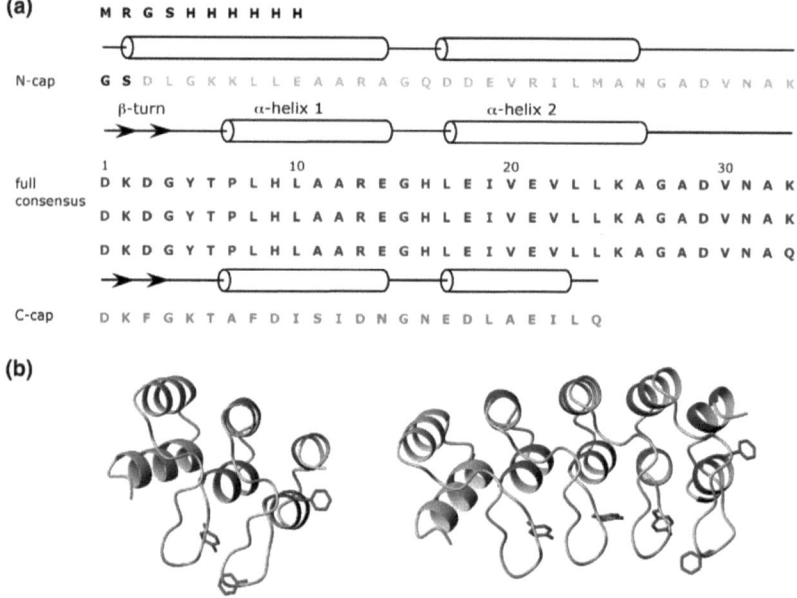

Fig. 1. (a) Amino acid sequence of NI₃C. The numbering of the internal consensus repeat is shown as used in the text. The N-terminal His tag, connected by a Gly-Ser linker, is shown in black. (b) Ribbon model representation (MOLMOL[4]) of NI₁C and NI₃C based on the structure of E3_5.[5] The colors are chosen to distinguish between the different repeats. The N-terminal repeat is depicted in orange, the internal consensus repeat is depicted in blue and the C-terminal repeat is depicted in green. The aromatic amino acids phenylalanine and tyrosine are represented as stick mode in red.

Pins can thus be considered as generalized examples for the study of AR protein folding. By characterizing the thermodynamic and kinetic parameters of these consensus proteins using circular dichroism (CD) and fluorescence spectroscopy, the dependence of stability, as well as of folding and unfolding rate constants, on repeat number was investigated. Moreover, we intended to gain mechanistic insight into the folding pathway of the three smallest proteins consisting of three to five AR repeats.

Results

Design of the consensus sequence

The "full-consensus" AR was based on a repeat module designed previously for a library of DARPins.[18] In the previous work, 7 out of 33 amino acids were allowed to vary in order to bind to target molecules. While the 26 fixed residues of the library repeat module were used without changes in the present study, defined residues had to be assigned to the six randomized potential interaction residues (positions 2, 3, 5, 13, 14 and 33) (Fig. 1a) and to the remaining randomized framework residue (position 26). For this purpose, we used the consensus analyses of sequence and structure described previously.[18] Positions 2 and 33 were defined as lysines, since this was the most frequent amino acid in these positions in our alignment (position 2: Lys, 19%; Arg, 13%; Asn, 13%; Ser, 12%; position 33: Lys, 21%; Arg, 16%; Trp and Asn, 9%). We chose glutamate for position 14 due to its most frequent occurrence (Glu, 22%; Asn, 13%). Even though arginine is not the most frequent amino acid in our alignment at position 13 (Gln, 22%; Lys, 19%; Arg, 18%), positively charged residues, when taken together, are most abundant. The first α-helices of each AR are in close proximity in juxtaposed repeats, and thus arginine should compensate for the negative charge of position 14. In addition, arginine has a high α-helical propensity.[19] The framework position 26 was defined as alanine, since this is the most frequently found residue in this position (40%) and since it could successfully be used repetitively for a previous full-consensus design.[20] The resulting consensus sequence is shown in Fig. 1a.

When comparing the full-consensus AR proteins investigated here with those described earlier,[20] the most important difference is the presence of capping repeats, which are identical with those we have published earlier[18] for our library. These appear to be essential for correct folding *in vivo* and full reversibility of folding *in vitro* at neutral pH.[21] Additionally, our sequence of the "internal" repeats is different in several positions from that of Mosavi *et al.*[20] In our numbering scheme[3,18] (cf. Fig. 1a), we start with the β-turn; thus, our position 3 corresponds to position 1 of the sequence from Mosavi *et al.* (see Supplementary Fig. 1).[20] At position 3, we have introduced Asp for reasons described above, while Mosavi *et al.* used Asn. Our chromophoric Tyr at position 5 was Arg in their work; at position 14, we used Glu instead of Asn for reasons described above. Four additional differences are in helix 2, which are "framework" positions already defined previously.[18] In position 19, Ile had been chosen (instead of Val) in our work because of its higher helical propensity; in position 21, Glu had been chosen (instead of Lys) because of its occurrence in the structure of GABPβ1; in position 22, Val had been chosen (instead of Leu) for the same reason; and, in position 25, Lys had been chosen (instead of Glu) for its opposite charge to Glu21.

The resulting protein designed here shows an almost perfectly alternating pattern of surface charges, as shown in the crystal structure of NI$_3$C,[22] and may help to partially explain the unusually high stability of these full-consensus AR proteins.

Construction, expression and characterization of full-consensus DARPins

The full-length proteins were termed NI$_1$C–NI$_6$C, where I represents the full-consensus repeat, the subscript represents the number of identical full-consensus repeat modules, and N and C correspond to the N- and C-terminal capping repeats (resulting in proteins with three to eight repeats in total). The capping repeats differ slightly from the full-consensus repeat in size and sequence (Fig. 1a) and have a hydrophilic surface. The C-capping repeat has been examined in detail.[21] The protein sequence of NI$_3$C and the model structure of NI$_1$C and NI$_3$C are shown in Fig. 1.

The six proteins were expressed in *Escherichia coli* in soluble form. Purification was done by immobilized metal-ion affinity chromatography. Gel filtration with multiangle light scattering [NI$_1$C: measured molecular mass, 12 kDa (expected molecular mass, 10.8 kDa); NI$_2$C: 14.9 kDa (14.4 kDa); NI$_3$C: 17.4 kDa (17.8 kDa); NI$_4$C: 22.4 kDa (21.4 kDa); NI$_5$C: 25.8 kDa (25 kDa); NI$_6$C: 27.8 kDa (28.5 kDa)] showed that the proteins were monomeric and had the correct size. SDS-PAGE analysis confirmed their purity, and mass spectrometry (data not shown) confirmed their correct composition and lack of degradation. The CD spectra of the six proteins are superimposable on the spectra of natural and other designed ankyrin proteins (data not shown).

Guanidine-hydrochloride-induced equilibrium unfolding

The denaturation of the six full-consensus DARPins NI$_1$C–NI$_6$C was monitored using both CD at 222 nm and tyrosine fluorescence. The stabilities increase with increasing repeat number, and this is observed both in guanidine hydrochloride (GdnHCl)-induced unfolding (Fig. 2) and in temperature-induced unfolding (Fig. 3). In GdnHCl-induced unfolding, we observe that only NI$_1$C, NI$_2$C and NI$_3$C can be unfolded (defined by the disappearance of the CD signal at 222 nm), while the longer pro-

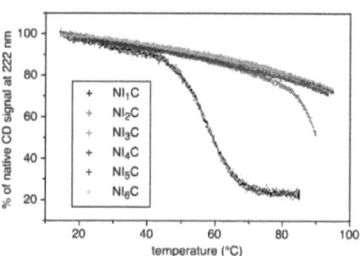

Fig. 3. Thermal melting of NI$_1$C–NI$_6$C followed by CD spectroscopy between 15 °C and 95 °C (see Materials and Methods). The derived midpoints of denaturation (melting temperature) are summarized in Table 1. The protein concentration was 10 µM. Melting was only 70% reversible for all proteins.

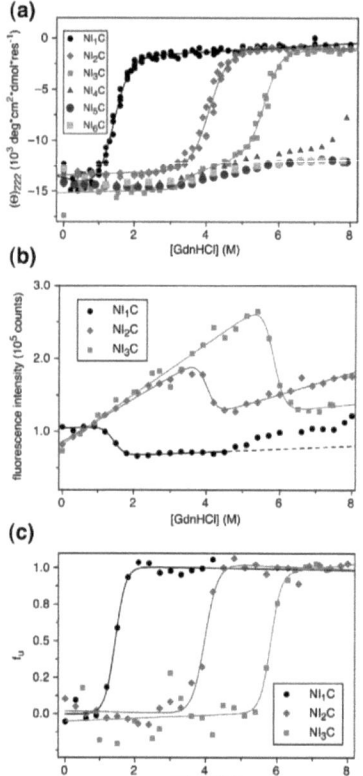

Fig. 2. GdnHCl-induced equilibrium unfolding of the six proteins NI$_1$C–NI$_6$C at 20 °C followed by (a) CD spectroscopy and (b and c) tyrosine fluorescence (b) plotted with fluorescence intensity units and (c) converted to fraction unfolded f_U (see Materials and Methods). The line represents the least-squares fit to a two-state model (except NI$_3$C, whose CD signal was fitted to a three-state model). The parameters of the fits are summarized in Table 1. The protein concentration was 10 µM.

teins (NI$_4$C–NI$_6$C) can only partially be unfolded. NI$_4$C reaches a major, but not completed, transition (loss of about 50% of the initial signal) at 8 M GdnHCl. All proteins from NI$_3$C to NI$_6$C show a small transition at around 4 M GdnHCl (see below).

The CD data, which show a single transition midpoint for NI$_1$C and NI$_2$C, can be described by a two-state model (Eq. (1)), which, of course, does not prove a fully cooperative system (see below). NI$_3$C, for which two transitions are visible, is best described by a sequential three-state model (Eqs. (3)–(9)). The first transition at 3.7 M GdnHCl would correspond to N = I, and the second transition I = U occurs at 5.6 M. This second CD transition coincides with the single fluorescence transition (Fig. 2b and c). We interpret this equilibrium intermediate as one in which the C-terminal capping repeat has unfolded[21] (see below). It appears that NI$_4$C, NI$_5$C and NI$_6$C also show this intermediate but do not denature further.

We also measured the GdnHCl-induced transitions by tyrosine fluorescence. Although the tyrosine residues are not in the hydrophobic core and the nature of the change in fluorescence signal during the unfolding transition is not completely clear, we could observe unfolding transitions with tyrosine fluorescence. The unfolding curves show a single transition (Fig. 2b and c) to lower fluorescence at about the same GdnHCl concentration as the main CD transition and a steep slope of the pre-transition and posttransition baselines (Fig. 2b) whose origin is currently not clear.

The thermodynamic parameters calculated for the only three proteins that can be fully unfolded, which are summarized in Table 1, were analyzed by a classical cooperative folding model. The m value measured for the almost globular NI$_1$C ($m = 2.6 \pm 0.2$ kcal mol^{-1} M^{-1}; Table 1) is consistent with the expectations derived from the buried surfaces[23] ($m_{\Delta ASA} = 2.7$ kcal mol^{-1} M^{-1}), while the m value determined for NI$_2$C is much smaller than would be expected for a typical globular protein of this size.

Results I

Table 1. Thermodynamic parameters for GdnHCl-induced and thermal unfolding of the proteins NI$_1$C, NI$_2$C and NI$_3$C

Protein	T_m [°C]	D_m [M]^a	Equilibrium data^b		Kinetic data^c	
			m [kcal mol^{-1} M^{-1}]	ΔG_0 [kcal mol^{-1}]	m [kcal mol^{-1} M^{-1}]	ΔG_0 [kcal mol^{-1}]
NI$_1$C	60	1.4	2.6±0.2	3.7±0.3	4.0±0.4	5.4±0.6
NI$_2$C	90	4.1	2.3±0.2	9.2±0.7	4.4±1.0	16.7±3.9
NI$_3$C^d	>100	3.7/5.6	1.7±0.9/3.0±0.6	19.7±4.6	–	–
NI$_4$C	>100	>8	≈8	–	–	–
NI$_5$C	>100	>8	–	–	–	–
NI$_6$C	>100	>8	–	–	–	–

^{a} Denaturation midpoint (M GdnHCl) at 20 °C.
^{b} Values obtained by two-state (NI$_1$C and NI$_2$C) or three-state (NI$_3$C) fitting.
^{c} Values obtained from kinetic three-state fit using $K^0 = ((k_N^0 k_{IU}^0)/(k_U^0 k_{IN}^0))$ and $m = (-m_{UI} + m_{IU} - m_{IN} + m_{NI})$.
^{d} Values for both transitions are given (for details, see the text).

This already suggests that, despite the fact that only one transition can be seen, the protein is not well described by a fully cooperative two-state model as typical for small proteins.

In a separate study,[21] we have shown experimentally that the small transition of NI$_3$C at about 4 M GdnHCl is consistent with a selective denaturation of the C-terminal capping repeat, whose different sequence, shorter C-terminal helix and fewer inter-repeat interactions may explain this phenomenon. Molecular dynamics calculations are also consistent with this interpretation.[21,24] In a different study, where consensus ARs were introduced between repeats 5 and 6 of the Notch AR domain, an equilibrium intermediate has been found as well. This intermediate disappeared when removing the C-terminal repeats 6 and 7.[25]

For the bigger proteins, evidence for some soluble aggregate formation was found at the highest GdnHCl concentration. This could be detected by measuring light scattering with a fluorimeter by recording the emission of light at 360 nm as a function of GdnHCl. Some onset of scattering was observed for NI$_4$C, NI$_5$C and NI$_6$C at 6–7 M GdnHCl, while no detectable aggregates were formed for NI$_3$C under these conditions (data not shown). NI$_3$C appeared to be fully denatured under these conditions (see Fig. 2a), and it is thus not the denatured species that partially aggregates. Taken together with the fact that the CD signal hardly changes for the larger molecules, this suggests that the unfolding of the C-terminal capping repeat no longer protects the molecules from some form of soluble aggregate formation at very high GdnHCl concentrations and that essentially native-like species (devoid of the solubilizing C-terminal capping repeat) start to associate under these conditions, somewhat reminiscent of a salting-out effect of native proteins. It is possible that these native-like species stabilize each other during this process (see below). Note that the design of even more stable C-terminal capping repeats has been successfully undertaken.[21]

Ising model describing the stability of full-consensus DARPins

Besides classical cooperative two-state or three-state folding models as described above, we have also analyzed the folding of the full-consensus DARPins by an Ising model. Briefly, this model does not assume a cooperative folding of the whole protein, but considers every repeat as an individual folding unit. The free energy of each repeat is assumed to be linearly dependent on denaturant concentration, characterized by an m value, while the interaction energy between neighboring repeats is considered to be constant, as long as both are folded. Similar models have been proposed previously for the study of repeat proteins.[11,26]

Because of the different stabilities of the capping repeats, we had to use a model that includes them as a separate unit with different ΔG_0 and m values. Even though we had found experimental evidence, triggered by molecular dynamics simulations,[21,24] that it is the C-terminal capping repeat that denatures at the lowest denaturant concentrations, we do not know whether the N-terminal capping repeat is also of somewhat lower stability than the internal repeats. To keep the model as simple as possible, we therefore considered both N-terminal and C-terminal capping repeats as having the same ΔG_0, which is lower than that of the internal repeats. This was done as the fits were significantly better (data not shown) than treating only the C-terminal capping repeats separately.

The Ising model, as used here, thus contains five parameters (see Eqs. (18)–(21) in Materials and Methods) describing all repeat proteins: the coupling energy J between repeats, the free energy of an isolated repeat ΔG_0, its denaturant dependence m, the free energy of isolated capping repeats $\Delta G_0'$ and their denaturant dependence m'. To determine these parameters, the equilibrium denaturation data of NI$_1$C, NI$_2$C and NI$_3$C, as measured by CD (Fig. 2), were globally fitted to Eq. (21), as shown in Fig. 4. This approach assumes that the CD signal is proportional to the number of folded repeats across the whole population, rather than, as in a two-state model, to the percentage of molecules with all repeats folded (for details, see Materials and Methods). The determined parameters of the Ising model are summarized in Table 2.

According to the Ising model, besides the completely folded and completely unfolded configurations, partially folded states are also populated. Figure 5a summarizes the most important possible

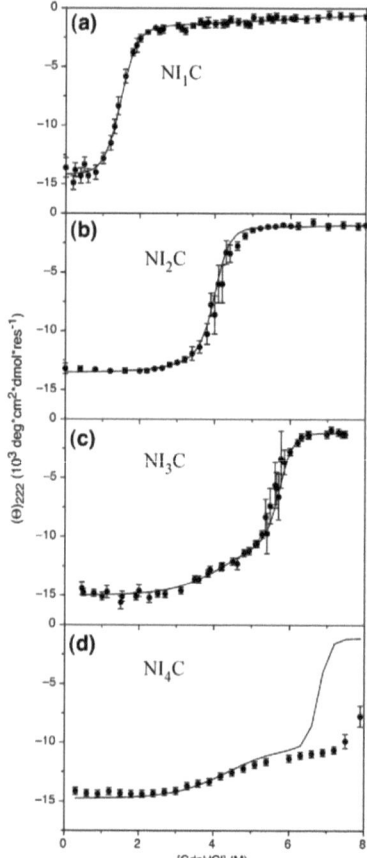

Fig. 4. Ising model fit of NI$_1$C, NI$_2$C and NI$_3$C data, and prediction of NI$_4$C. Equilibrium GdnHCl-induced unfolding of the three proteins NI$_1$C (a), NI$_2$C (b) and NI$_3$C (c) at 20 °C followed by CD spectroscopy. The line represents the fit to the Ising model (see Materials and Methods). The parameters are summarized in Table 2. (d) Prediction using the Ising model fit and experimental data for NI$_4$C.

repeat folded are those maximally unstable at any denaturant concentration, as the free energy of folding only one repeat is always unfavorable, and no stabilization from interaction with the neighbors can be provided. Indeed, isolated single ARs were experimentally found not to be stable.[20] This state with exactly one repeat folded thereby separates the two regions of minimal free energy, corresponding to the unfolded states on one side and the folded states on the other side (Fig. 5b).

At zero (or low) denaturant, the most stable state is, as expected, the one in which all repeats are folded. At high denaturant, for the three proteins shown here, the most stable state is the one in which all repeats are denatured, also as expected. However, at high denaturant concentration, the most stable among the native states are those with one and/or both unfolded terminal repeats but folded internal repeats. Thus, under highly denaturing conditions, these states become more stable than the completely folded configuration for all the DARPins examined with this model (Fig. 5c and d). In other words, the model predicts a denaturant-dependent shift of the most stable state in the native free-energy basin. This behavior results in an overall nonlinear dependence of the stability $\Delta G_L(D)$ on [D], that is, the free-energy difference between the most stable conformer in the native basin and the unfolded state (Fig. 6). The kinks observed in $\Delta G_L(D)$ are due to the abovementioned shift in the native state of the protein as [D] increases.

With the parameters determined from the fit of the equilibrium denaturation data measured by CD for NI$_1$C–NI$_3$C, it is possible to extrapolate the Ising model to describe the behavior of the larger proteins NI$_4$C–NI$_6$C (Figs. 4d and 6), which can no longer be experimentally unfolded (except by heating in high GdnHCl; Fig. 7). The slope of the stabilities $\Delta G_L(D)$ in proximity to the predicted transition midpoint (intersection with the x-axis of Fig. 6) provides m_L values (Table 3). We report these to allow a comparison with standard two-state fits of the data (even if these are only possible for NI$_1$C and NI$_2$C) (Table 1). In Table 3, predicted transition midpoints and stabilities at 0 M GdnHCl are also reported. In the Ising model, the stability at 0 M GdnHCl depends linearly on the number of repeats L. In the homogeneous Ising model (all repeats are identical; not shown here), the m_L value is also linear with the number of repeats. Because of the presence of terminal repeats of different intrinsic stabilities in the model presented here, m_L tends to be linear only for large L ($L \geq 5$), where the relative influence of the terminal repeats becomes smaller.

states for NI$_1$C, NI$_2$C and NI$_3$C, where one to four repeats are folded. Note that any folded repeats will always be adjacent, as the population of other states is insignificant. In addition, a lower free energy is obtained if the folded states are all internal than if they include a capping repeat. The free energy of the states of the protein as a function of the GdnHCl concentration [D] (Fig. 5b–d) provides a coherent interpretation of the experimental data. According to this model, the configurations with only one

Table 2. Parameters obtained by fitting the equilibrium data of NI$_1$C, NI$_2$C and NI$_3$C to the Ising model (see Fig. 4)

J	−14.2±0.7 kcal mol^{-1}
ΔG_0	3.3±0.2 kcal mol^{-1}
m	1.1±0.1 kcal mol^{-1} M^{-1}
$\Delta G_0'$	10.6±0.6 kcal mol^{-1}
m'	0.83±0.04 kcal mol^{-1} M^{-1}

Results I

Folding and Unfolding of Ankyrin Repeat Proteins

A direct comparison of model predictions with experimental data for NI$_4$C (Fig. 4d) and larger constructs (Table 3) shows that the model apparently underestimates the stability of these proteins. This led us to propose a possible hypothesis for the discrepancy. The long constructs, at high denaturant concentration, have unfolded terminal repeats, according to the Ising model, but are otherwise still essentially folded and may be stacking together due to the now-exposed hydrophobic surfaces at the edge of the consensus repeats. These elongated complexes are more stable than the monomeric molecule because of their larger effective length. As reported above, light-scattering experiments confirmed that the DARPins NI$_4$C–NI$_6$C begin to form soluble aggregates at high denaturant concentration. The presence of such soluble, essentially native-like, aggregates might explain the discrepancy between the predictions from the Ising model and the experimental data for the large constructs. Further experiments outside the scope of the present work will be needed to confirm the exact nature of the species giving rise to increased light scattering at very high GdnHCl concentrations.

Thermal stability

The thermal unfolding of all six proteins was monitored by the change in CD signal at 222 nm (Fig. 3). The NI$_1$C transition is sigmoidal with a midpoint $T_m = 60$ °C. Melting of NI$_2$C was not complete, but T_m can be estimated to be 90 °C (Table 1). The bigger proteins, NI$_3$C–NI$_6$C, could not be thermally unfolded at all. Melting was only possible by heating them in 5 M GdnHCl or by heating them in a buffer of pH 3.5 (data not shown). Only by using these unusually strong denaturing conditions could the proteins NI$_3$C, NI$_4$C, NI$_5$C and NI$_6$C be unfolded (Fig. 7). For NI$_3$C and NI$_4$C, the pretransition

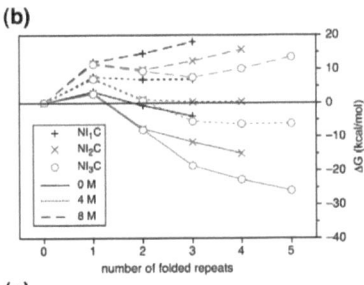

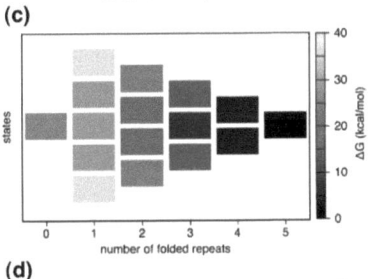

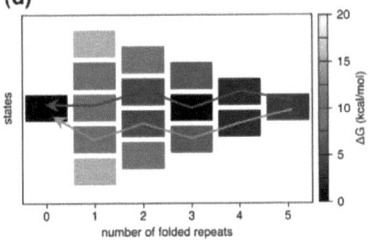

Fig. 5. Possible states and average free energy of the states of NI$_1$C, NI$_2$C and NI$_3$C. (a) Cartoon of possible configurations in NI$_1$C, NI$_2$C and NI$_3$C, according to the Ising model. Folded (●) and unfolded repeats (○) are displayed with different symbols. For states that contain more than one folded repeat, only those in which folded repeats are adjacent to each other are shown (and counted here). For example, for NI$_1$C, the state (●-○-●) is hardly populated, such that it can be neglected next to (○-●-●) and (●-●-○), explaining the number "2." Since the folded internal repeats have a different energy from the folded terminal ones, their respective numbers are separated by a (+) sign. For example, in NI$_1$C, there is only one state with a folded internal repeat (○-●-○), while there are two with terminal repeats (●-○-○) and (○-○-●), explaining the numbers "1+2." (b) The free energies of NI$_1$C, NI$_2$C and NI$_3$C as a function of the number of folded repeats plotted at three different GdnHCl concentrations. Free energies obtained at 0 M GdnHCl (solid lines), 4 M GdnHCl (short dashed lines) and 8 M GdnHCl (long dashed lines) are plotted. (c and d) Free energy of NI$_3$C considering only states with contiguous folded repeats at 0 M GdnHCl (c) and 6 M GdnHCl (d). Each box represents a state, and its color indicates its free energy at the GdnHCl concentration of the plot. States with the same number of contiguous folded repeats are arranged vertically in the figure and ordered, from bottom to top, such that the configurations with folded repeats that include the N-terminal repeat are represented by the lower boxes, those with a C-terminal folded repeat are represented by the upper boxes and those with only consensus (middle) repeats folded are represented by the middle boxes. At 0 M GdnHCl (c), as expected, the state with all the repeats folded is the most stable, while at 6 M (d), the most stable state is the completely unfolded one, but another one that is almost as stable coexists in the native free-energy basin, which has three central repeats folded and the terminal repeats unfolded. The blue pathway contributes to the slowest phase, while the green pathway may give rise to a faster phase. All of the pathways have to progressively cross the diagram from right to left, losing a single (i.e., not more than one) folded repeat at every step.

Results I

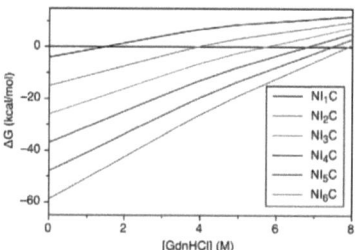

Fig. 6. Stability as a function of GdnHCl for the six consensus DARPins of different lengths, as obtained by the Ising model. Stability is defined as the energy difference between the lowest energy state among the folded ones (at the corresponding denaturant concentration) and the lowest energy state among the unfolded ones, which is the state with all repeats unfolded. This quantity plotted equals the difference between the activation free energy of unfolding and the activation free energy of folding, as determined from the barrier heights in Fig. 5b.

baselines start at values that are significantly below the value of 100% native protein, as these proteins are already partially denatured in 5 M GdnHCl at room temperature (see also Fig. 2a). The transition midpoints T_m in 5 M GdnHCl were approximately 56 °C for NI$_3$C, 76 °C for NI$_4$C, 88 °C for NI$_5$C and approximately 96 °C for NI$_6$C. The melting was only 70% reversible, as judged from the recovery of the CD signal after cooling, and the thermal denaturation could thus not be used to derive ΔG values. This experiment underlines the extraordinary stability of the proteins. We believe that the consensus design,[18] which has "idealized" the protein structure by introducing many critical interactions in every single repeat, combined with the choice of the residues in the previously randomized positions (see above), which lead to a "checkerboard" arrangement of charges,[22] with maximization of attractive interactions, contributes to this high stability.

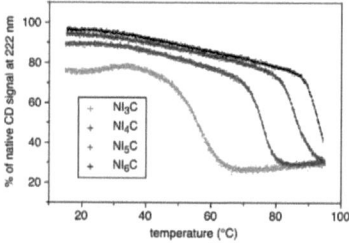

Fig. 7. Thermal melting of NI$_3$C, NI$_4$C, NI$_5$C and NI$_6$C in the presence of 5 M GdnHCl. The proteins NI$_4$C–NI$_6$C can only be denatured by elevated temperatures in 5 M GdnHCl. The stabilities of NI$_4$C, NI$_5$C and NI$_6$C can be distinguished qualitatively, but due to nonreversibility, no thermodynamic parameters can be calculated.

Table 3. Determined and extrapolated $D_{m,L}$, m_L and ΔG_L^0 values for the DARPins investigated in the Ising model

DARPins	$D_{m,L}$ [M][a]	$m_L(D_{m,L})$ [kcal mol^{-1} M^{-1}][b]	ΔG_L^0 [kcal mol^{-1}][c]
NI$_1$C[d]	1.47±0.08	2.7±0.2	−4.03±0.54
NI$_2$C[d]	3.96±0.07	2.8±0.2	−15±0.8
NI$_3$C[d]	5.56±0.18	1.9±0.8	−25.9±1.1
NI$_4$C[e]	6.65±0.1	2.4±0.7	−36.9±1.4
NI$_5$C[e]	7.35±0.08	3.3±0.95	−47.8±1.6
NI$_6$C[e]	7.8±0.08	4.3±0.6	−58.8±1.9

[a] Denaturation midpoint (M GdnHCl).
[b] Slope of ΔG_L at the transition midpoint.
[c] ΔG_L at 0 M GdnHCl.
[d] Experimentally determined.
[e] Extrapolated as described in the text.

The experiment in Fig. 7 also shows that the distances between the T_m values of NI$_3$C–NI$_6$C decrease with increasing repeat number. A dependence of stability on length has also been observed for other consensus repeat proteins[25,27] and dissected versions of the natural *Drosophila* Notch receptor.[11] However, such an enormous stability increase has not been observed yet.

Unfolding and refolding kinetics of NI$_1$C, NI$_2$C and NI$_3$C

Since only the shorter AR consensus proteins could be denatured at room temperature, the folding and unfolding kinetics of the DARPins NI$_1$C, NI$_2$C and NI$_3$C were studied as a function of denaturant concentration and are represented as chevron plots. The folding and unfolding rates were determined by monitoring the changes in ellipticity at 225 nm after dilution out of, or into, GdnHCl at 20 °C using a stopped-flow instrument. Slow reaction phases (>100 s) were recorded by manual mixing.

It is immediately apparent that the folding rates are equally fast for all proteins, while the unfolding rates are slow and get slower with increasing numbers of repeats (Fig. 8a).

All kinetic traces of NI$_1$C (folding and unfolding) could be fitted to a single exponential function (Eq. (10)). While NI$_2$C refolding was monophasic, NI$_2$C unfolding was fitted to a double exponential function (Eq. (11)) representing two unfolding phases, with amplitudes decreasing from 70% to 10% with GdnHCl concentration for the slow phase and increasing from 30% to 90% for the fast phase. Both unfolding phases were recorded after fast mixing, and the slower unfolding phase was also confirmed by manual mixing. NI$_3$C refolding was monophasic, too, while the unfolding reaction could be divided into three phases (triple exponential). The slowest unfolding phase was recorded by manual mixing (60% of the total amplitude), while both fast unfolding phases were recorded by fast mixing (together 40% of the total amplitude: 30% for the faster phase and 10% for the slower phase). Representative kinetic traces of NI$_2$C and NI$_3$C are shown in Supplementary Fig. 2. In addition to the

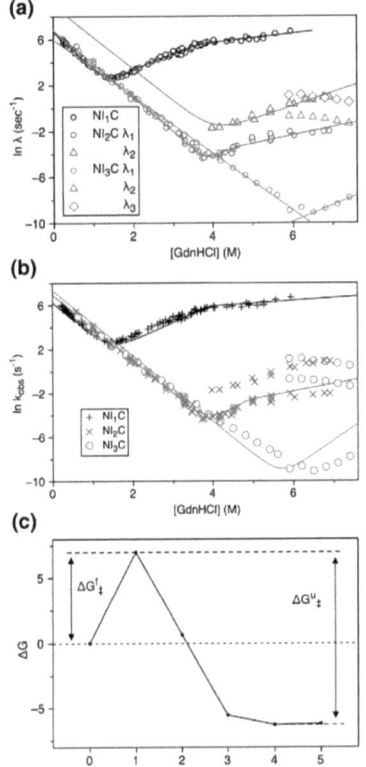

Fig. 8. Chevron plots of NI_1C, NI_2C and NI_3C, and prediction of the kinetics by the Ising model. (a) Experimental chevron plots of NI_1C, NI_2C and NI_3C at 20 °C, followed by CD spectroscopy. The left limb corresponds to the GdnHCl-dependent rate constants for refolding, and the right limbs correspond to the unfolding rate constants. The lines represent the best fit to the data using the kinetic on-pathway intermediate three-state model.[28] The rate constants are summarized in Table 4. (b) Theoretical chevron plots of NI_1C, NI_2C and NI_3C (lines) obtained by using the height of the folding/unfolding barriers from Fig. 5b to estimate the folding and unfolding rates and their dependence on GdnHCl and protein length. The preexponential factor a has been chosen as $a = 1.3 \times 10^5$ s^{-1}exp(−1.15 [D]) to best match the experimental data in the chevron plot [superimposed in the figure as symbols; same data and symbols as in (a)]. (c) Graphical explanation of the procedure used to obtain the height of the folding or unfolding barriers using the data from Fig. 5b. We show, as an example, the NI_3C free-energy profile at 4 M GdnHCl, where the activation free energy for folding is 7 kcal mol^{-1}. Substituting this number into the formula for the rate $k = a$ exp(−$\Delta G_\ddagger / kT$) with the preexponential factor a at 4 M GdnHCl, we obtain the rate that was then plotted in (b) for NI_3C at 4 M GdnHCl.

CD data discussed above, folding kinetics have also been measured by tyrosine fluorescence (data not shown). While folding rates were found to be very similar, the fastest unfolding phase of NI_3C was 3-fold faster, suggesting that, in this case, loss of tertiary structure precedes loss of secondary structure. In the unfolding of NI_2C, however, all rates measured by CD and tyrosine fluorescence were found to be very similar.

Plotting the fitted rates as a function of denaturant concentration revealed clear deviations from the classical V-shaped two-state chevron plot for all proteins. NI_1C unfolding and refolding are monophasic, but the unfolding limb exhibits a curvature (Fig. 8a). This curvature is even more pronounced at a lower temperature (5 °C; data not shown). One way to explain this behavior in a classical framework is to assume a sequential three-state model with a metastable high-energy intermediate.[29] In this model, the metastable intermediate is not populated enough to cause a second observable folding or unfolding phase, but it does influence the denaturant dependence of the unfolding rate. Such a behavior can be seen in the nonlinear slope of the unfolding limb of the NI_1C chevron plot. Another curved chevron plot has also been found for the ankyrin protein myotrophin[16] that consists of one repeat more than the DARPin NI_1C.

NI_2C kinetics are monophasic in the refolding limb, while the unfolding reaction is best described by a double exponential (Supplementary Fig. 2). The NI_2C unfolding limb of the slower unfolding phase shows a curvature, too. The NI_2C chevron plot could be well fitted by a sequential three-state model (Eq. (12)), where the unfolding intermediate state is more populated than for NI_1C and appears to be formed before reaching the fully unfolded state (on-pathway intermediate). Therefore, a second kinetic unfolding phase is observed (Fig. 8a).

Two unfolding phases were also found for the Notch ankyrin domain with seven ARs.[13] However, the Notch ankyrin domain also reveals a second refolding phase, which has been attributed to prolyl isomerization in the unfolded state, and that protein is much less stable and folds much more slowly than the DARPin NI_2C, which is three repeats shorter and has only a single folding phase. The Notch domain Nank1-7Δ has four *trans*-prolines and one disordered proline, while the tumor-suppressor protein p19 has even seven *trans*-prolines. In contrast, NI_1C has one *trans*-proline, NI_2C has two *trans*-prolines and NI_3C has only three *trans*-prolines. The bigger number of proline residues in the Notch domains might explain why slow folding phases were detectable, while they were not detectable in our case.

The kinetics of NI_3C are more complex: the refolding reaction is again monophasic as observed for NI_1C and NI_2C, while unfolding was separated into three phases (two phases monitored by stopped-flow mixing, one phase observed by manual mixing). A three-state model is not able to describe such complex kinetics, and models with more than three states are very difficult to apply

Table 4. Unfolding and refolding rates of NI_1C, NI_2C and NI_3C

DARPins	k^0_{Uf}	k^0_{fU}	k^0_{fN}	k^0_{Nf}	k^0_u	k^0_f
NI_1C	776±43	3.6±1.4	3589±3417	66±17	3.4	638
NI_2C	892±103	$1.1 \times 10^{-3} \pm 7 \times 10^{-4}$	6961±4.6×10⁴	$1.4 \times 10^{-3} \pm 7 \times 10^{-4}$	6.3×10^{-4}	790
NI_3C	n.d.	n.d.	n.d.	n.d.	1×10^{-8}	400

The rates (s^{-1}) at 20 °C, extrapolated to native buffer, were obtained by fitting a sequential three-state model with on-pathway intermediate.[28] k^0_f (Eq. (16)) and k^0_u (Eq. (17)) of NI_1C and NI_2C were calculated from the fitting parameters k^0_{Uf}, k^0_{fU}, k^0_{fN} and k^0_{Nf}.

quantitatively as they exhibit too many parameters, making a converging fit very difficult.

The kinetic data thus indicate a more complex model than the two-state model for NI_1C and NI_2C. In contrast, the equilibrium data can be fitted well to a two-state model, and there was thus no hint for a deviation from that fit alone. We therefore also report the ΔG_0 and m values calculated from kinetics (Table 1) using the four rate constants k_{ij} according to the sequential three-state model. A global fit to both kinetic and thermodynamic data of NI_1C and NI_2C using the three-state model (with metastable and stable intermediates, respectively), which fits the data well, gives values rather similar (data not shown) to the ΔG_0 values derived from kinetics alone (Table 1). We assume that this similarity is largely an effect of the kinetics dominating the parameter fit, since equilibrium measurements offer looser constraints on the parameters of the model than kinetic data because of the additional degrees of freedom provided by the choice of the baselines. However, even the kinetics themselves do not fully determine all the parameters of the three-state model, as some of them are also rather insensitive to the global fit, as can be seen by the size of some of the error margins in Table 4. In summary, when incorporating all observations, these proteins are not consistent with a two-state system. Instead, their folding behavior is consistent with three-state models, even if some of their parameters remain rather poorly constrained, resulting in significant error margins.

The kinetic parameters of NI_1C, NI_2C and NI_3C are summarized in Table 4. The first important observation is that the folding rates do not vary much between all three proteins. Folding is monophasic, and the folding process thus needs to cross essentially the same transition state, independent of the number of repeats in the protein. This is intuitively appealing, as it suggests that a very small number of repeats, in the simplest case one, must be folded in the transition state, which is common to all proteins investigated here.

In sharp contrast, the unfolding rates decrease tremendously (a factor of 10^4 from NI_1C to NI_2C and from NI_2C to NI_3C) with increasing repeat number. In order to compare the folding and unfolding rates of the three proteins, the refolding rate and the slowest unfolding rate of NI_3C were used. With the longer proteins, this unfolding rate could not be measured, as they do not fully denature at all under these conditions, as explained above. If the slowest unfolding rates are considered, unfolding rates of 0.6 s^{-1}, 1.3×10^{-8} s^{-1} and 1.2×10^{-16} s^{-1} (corresponding to a half-life of 1.8×10^8 years) are extrapolated using the Ising model (see Folding and Unfolding Pathways According to the Ising Model) to zero denaturant for NI_1C, NI_2C and NI_3C, respectively. The extrapolated values for NI_4C, NI_5C and NI_6C would even be predicted to be a factor of 10^8 slower for each repeat added, but, of course, we have no possibility to test the validity of these extrapolations at this point. The stability is thus directly reflected in their (predicted) half-lives in native buffer. As the folding rate is certainly not slow and does not vary between the three proteins, the very slow denaturation rates are thus a direct reflection of the very low free energy of the folded state, which raises the barrier to unfolding correspondingly.

Folding and unfolding pathways according to the Ising model

In experiments probing the kinetics of folding and unfolding, the proteins are expected to follow the minimum free-energy pathways, connecting the initial state with the equilibrium state under the new conditions (i.e., under different denaturant concentrations). The Ising model allows for the reduction of a complex protein folding problem into a simple reaction occurring between a finite set of states. In this simplified conformational space, the slowest relaxation rate is determined by the largest free-energy barrier found along the minimum free-energy pathway connecting the initial state and the final state. Given a number of folded repeats, the Ising model assigns the minimum free energy (i.e., the maximal probability of being occupied as computed by Eq. (19)) to the state where the folded repeats are contiguous. Assuming that the repeats fold and unfold one at a time, according to the Ising model, the largest barrier between the fully folded state and the fully unfolded state corresponds to the state with a single repeat folded (Fig. 5b). From the height of the free-energy barrier separating the folded state from the unfolded state, it is possible to obtain folding and unfolding rates by providing a valid preexponential factor† (defined by an intercept and denaturant dependence) (Fig. 8c). In the present case, by choosing a preexponential

---

† The preexponential factor is the parameter a in the formula connecting the folding rate to the activation free energy: $k = a \exp(-\Delta G_\ddagger/kT)$.

factor of $a = 1.3 \times 10^5$ s^{-1} exp(−1.15[D]), it is possible to draw a chevron plot for the slowest relaxation rates, using the height of the folding/unfolding barriers in Fig. 5b as activation free energies $\Delta G_\ddagger$ (Fig. 8b). The model captures the bending of the unfolding arm of NI$_1$C and NI$_2$C. The distance between the unfolding arms of the different proteins is also predicted quite accurately.

It may be worth underlining that this is not a conventional fit of the slowest kinetic traces of the proteins. Here, only the data from the equilibrium denaturation curves have been fitted using the Ising model, and the latter has been used to derive the kinetic properties using no additional parameters apart from the preexponential factor and its denaturant dependence. The preexponential factor, in this case, includes information on the rate of structure formation/structure loss of a single repeat. This term depends on denaturant concentration. This information is not present in the equilibrium data and thus has to be provided separately. The two parameters describing the preexponential factor are the minimal set of data needed to translate the height of the free-energy barriers of the Ising model into rates. Indeed, visually, the choice of the two parameters defining the preexponential factor is made by superimposing a straight segment of the folding arm of one of the theoretical chevron plots to the corresponding experimental data by suitably adjusting these two parameters. All the other features of all the chevron plots are then fixed. That means that the transition midpoints, the kinetic m_U values and the location (on the [GdnHCl] axis) and size of the change in m values of *all* the chevron plots are fixed once the folding arm of a *single* chevron plot is fixed. The Ising model preexponential factor, which reports the rate of structure formation/structure loss of the single repeat, at 0 M GdnHCl provides a lower boundary for the classical folding preexponential factor known from the literature (p. 558 of Fersht[30]), which reports the speed of diffusion at the top of the free-energy barrier.

The discrepancies found with the experimental kinetic data for NI$_3$C at high denaturant concentration might be related again to transient "salting-out" phenomena that slow the complete unfolding of the protein. The high consistency of data interpretation provided by the model is summarized below.

Discussion

The folding behavior of a series of six full-consensus DARPins has been analyzed by experimental thermodynamic and kinetic experiments. We examined the dependence of stability on repeat number, as well as the correlation between the folding and unfolding rates and the repeat number. Moreover, we propose a description of the folding mechanism of the three full-consensus DARPins NI$_1$C, NI$_2$C and NI$_3$C using two different models, which also allow us to predict the behavior of other members of this series of proteins.

Consensus DARPins are very stable proteins

The first observation is that our consensus design yielded extremely stable proteins. Equilibrium experiments showed that adding repeats increases the thermodynamic stability of the proteins, and, with NI$_6$C, an ankyrin protein with eight repeats, we still have not reached a stability limit, as judged from the distances between the T_m in 5 M GdnHCl. The proteins NI$_4$C, NI$_5$C and NI$_6$C, which are not completely denaturing as judged from their CD signal but show an increase in light scattering at 6–7 M GdnHCl, may associate via the exposed interface after one or both caps have denatured.[21] This soluble aggregate state of quasi-native proteins may provide the proteins with an even larger stability than the monomeric state. Experimentally, neither high temperature nor 8 M GdnHCl is able to unfold them. Only by combining heat and GdnHCl could we unfold these molecules and distinguish their stabilities. The stability increases with increasing repeat number; however, due to the extremely high thermodynamic stability and the inability to fully denature the larger proteins (Figs. 2 and 7), it is difficult to quantitate this for the whole series.

In Fig. 7, it is apparent that the distances between the main transition midpoints become smaller with increasing repeat number. This raised the question as to whether we can calculate the equilibrium parameters for NI$_4$C, NI$_5$C and NI$_6$C, as well as a maximal asymptotic denaturation midpoint of DARPins NI$_x$C with large x. Calculations using the Ising model provide theoretical values for the large DARPins (for NI$_4$C, NI$_5$C and NI$_6$C, see Table 2). However, these values, being the product of an extrapolation process, should be considered as only indicative. The assumed linearity of both stabilities and m_L for large repeat numbers L (see Results) leads to an asymptotic value for the transition midpoint of the repeat proteins. In the case of the DARPins, this value (∼11 M GdnHCl) is well above the experimentally accessible range.

Unfolding rates decrease enormously with increasing repeat number

Second, we analyzed the dependence of the folding and unfolding rates on the repeat number. The dependence of the stability on the number of repeats is thus determined by a change in the unfolding rate, while the refolding rates of all three proteins are very similar (Table 4). At the molecular level, we imagine the following: in a molecule with many repeats, repeat interactions broken during unfolding by fluctuations will usually reseal and reanneal, since the remaining folded fragments do not denature. Only if the remaining folded fragment is small (consisting of one or two repeats) will it denature. The more repeats there are in the molecule, therefore, the less likely it will generate folded fragments that are small enough to continue complete denaturation. In the case of NI$_3$C, we have two strong interactions by the three highly compatible full-consensus interfaces (Fig. 1); in NI$_1$C, there is none; and, in NI$_2$C, there is only one such strong interaction. This explains the decrease of k_u with repeat number and the very slow unfolding rate of NI$_3$C.

In the language of Φ-value formalism,[30] the addition of repeats would be assigned a formal value of 0 for the folding reaction, as this addition would not change the transition state in folding but would only contribute interactions on the native side of the transition state. This formalism only conceptualizes the role of the additional repeats in stabilizing the native state, but of course makes no statement about the nature of the transition state within a single repeat.

Interpretation using classical cooperative folding models

None of the studied proteins can be interpreted using the classical two-state model of protein folding as the chevron plots exhibit nonlinearity in the unfolding limbs and, for the larger proteins, a multiphasic behavior. This can be treated by a formalism,[28] according to which folding proceeds through transiently populated and partially folded intermediates that are separated by major free-energy barriers along a reaction coordinate. The intermediates can be considered as local minima between the global minimum of the native state and the ensemble of unfolded states. The simplest case for such complex kinetics is the sequential three-state mechanism with folding through an on–pathway high-energy intermediate.[29] When several kinetic phases can be detected, the intermediate state is more populated. In our case, we have three states: N, U and an on-pathway intermediate I.[28]

The rates of NI$_1$C are all monoexponential, yet due to the curvature in the unfolding limb, we have to assume a deviation from a simple two-state model for this protein as well. Indeed, the data can be fitted using a sequential three-state mechanism with high-energy intermediate. Consistent with these results, differential scanning calorimetry experiments revealed a deviation of an NX$_1$C library member from a two-state model at a higher temperature, while in the kinetic unfolding at 5 °C, no intermediates were detected.[31] NI$_2$C can also well be fitted to the on-pathway three-state mechanism, with the intermediate being more populated than in NI$_1$C. The kinetics of NI$_3$C, however, are even more complex as we detected three separated phases for unfolding. This can no longer be quantitatively fitted, as there would be too many parameters.

Interpretation using an Ising model

According to a second model, based on the classical one-dimensional Ising chain, the DARPin constituent repeats can fold independently and interact via nearest-neighbor coupling. The model provides a very good fit to the equilibrium denaturation data (Fig. 4) and an appealing rationalization of the kinetic experiments. Namely, the model fitted only on equilibrium data offers a prediction for the free energy of partially unfolded states (Fig. 5) that allows for the drawing of a chevron plot for each DARPin (Fig. 8b). The theoretical chevron plot is in good agreement with the experimental data and provides a validation for the Ising model of the DARPins. This model, in fact, provides a comprehensive explanation for some experimental data summarized in the following.

The Ising model fitted to DARPins predicts that the activation free energy of folding is practically the same for all the proteins, irrespective of their length (Fig. 5b): after the nucleation step, provided by the folding of one of the consensus repeats, the other repeats condensate on this folding nucleus. In the same way, according to the Ising model, the activation free energy of unfolding increases with the number of repeats because any folded repeat added to the folding nucleus provides a further decrease in the free energy of the native state of the molecule (Fig. 6). The kink in the slowest unfolding limb of the chevron plot of NI$_1$C and NI$_2$C is ascribed to the stabilization of partially unfolded states, with one (for NI$_1$C) or two (for NI$_2$C) terminal repeats unfolded (Fig. 5b). In other words, while the transition state remains the same as [D] increases, the most stable conformer in the native free-energy basin changes with [D], causing changes in the major unfolding free-energy barrier.

In more detail, the appearance of multiple phases in the unfolding arm of the chevron plots of NI$_2$C and NI$_3$C could be explained by several mechanisms. One possibility that is very plausible in the classical model is that the fast phases may correspond to the fast unfolding of the less stable terminal repeats at high GdnHCl concentration. This rate is limited by the intrinsic rate of the structure disruption of the terminal repeats. A further splitting of the observable phases may be caused by the different natures of the N- and C-terminal repeats.[21]

In principle, multiple pathways may exist from the fully folded state to the fully unfolded state, some of which may not reach the lowest energy intermediate with the unfolded termini. It would be difficult, however, to determine the relative amplitudes of such pathways using the classical model.

In contrast, in the Ising model, we obtain an energy landscape as a function of GdnHCl. This landscape predicts an intermediate with the C-cap and the N-cap unfolded at 6 M GdnHCl for NI$_3$C (Fig. 5d). The slowest of the three unfolding phases of NI$_3$C at 6 M GdnHCl (Fig. 5d) has been thus associated with the pathways passing through these most stable intermediates (three central repeats folded; state in dark brown), while the two fast phases may correspond to the pathways passing through the less stable intermediates (two with three repeats folded, including a terminal one). We illustrate this in Fig. 5d.

Figure 5c shows a similar pattern as in the free-energy landscape described by Mello and Barrick, where the fully folded state and the fully unfolded state are free-energy minima, the states with only one folded repeat have the highest (most unfavorable) free energy and the other states have a free energy that decreases roughly linearly with the number of folded repeats.[11] In this work, using full-consensus designed proteins with identical internal repeats, we have, however, removed most of the heterogeneity present in the study of Mello and Barrick.[11] As a consequence, we have a large degree of degeneracy due to the presence of many states with the same free energy.

Comparison to the folding of other proteins

Comparing the rates with the natural AR proteins p16 and p19,[6,9] our DARPins fold with faster rates and unfold much more slowly at 20 °C.

Rates for another class of consensus repeat proteins have been measured for tetratricopeptide repeat (TPR)[27] and a hybrid ankyrin construct[25] consisting of natural and consensus ARs. Our result—unfolding rates slowing with the number of repeats —is in agreement with the kinetic study of Main et al.[27] In their study, the unfolding rate decreased by a factor of 36 when comparing the two-repeat CTPR2 with the three-repeat construct CTPR3 (0.35 s^{-1} and 0.01 s^{-1} for the two-repeat and three-repeat constructs, respectively),[27] while the folding rates were similar ($\approx 20,000$ s^{-1} and $35,000$ s^{-1} for CTPR2 and CTPR3, respectively). Compared to DARPins, these proteins unfold extremely fast (10^2- to 10^7-fold faster than the DARPins) and show the classical two-state behavior; therefore, possible intermediate states cannot be detected with standard experimental methods. In another study using an Ising model to describe a series of consensus TPR proteins,[26] a stabilizing energy of about 4 kcal mol^{-1} per repeat was obtained, while our DARPins gained 11 kcal mol^{-1} in stability per additional repeat (see Fig. 6). This illustrates why adding ARs decreases the unfolding rate much more than the addition of TPR.

A theoretical study by Plaxco et al. has suggested that there may be a linear correlation between the folding rate and the topology [expressed as relative contact order (RCO), the contact order normalized by the length of the protein] of small single-domain proteins exhibiting two-state kinetics.[32] Proteins with low RCO (i.e., containing mainly residues interacting with other residues that are close in sequence distance) fold very fast. Proteins with high RCO fold more slowly. In repeat proteins, which exclusively exhibit local interactions, the RCO does not change with increasing repeat number, and thus no change in the folding rate would be expected either. The folding rates‡ predicted from this model of NI$_1$C, NI$_2$C and NI$_3$C are 8.9×10^6 s^{-1}, 9.3×10^6 s^{-1} and 9.4×10^6 s^{-1}, respectively, while the corresponding measured folding rates at 20 °C range from 450 s^{-1} to 800 s^{-1} (638 s^{-1}, 790 s^{-1} and 450 s^{-1}, respectively; see also Table 4). Although the predicted folding rate is 10^4 times faster than the experimentally observed rates and is thus not in good quantitative agreement, it is interesting to see that the folding rates of all three proteins are very similar to each other, consistent with the model.

The Ising model suggests that folding and unfolding occur along multiple pathways following a nucleation–condensation mechanism, where multiple nucleation sites are possible. The fitting of the data to the model also suggested a possible size for the nucleation site (one consensus repeat for DARPins) and supports the emergence of stable partially folded intermediate states at high GdnHCl concentration.

Conclusions

Our study revealed two major insights: even though many AR proteins show two-state behavior in equilibrium studies, all kinetic studies performed with AR proteins to date have proven that the folding mechanism is more complex, with at least one intermediate state. In addition, our generalized full-consensus DARPin series confirms these findings. Second, the stability of the repeat proteins, characterized by short-range interactions and low contact order, is determined by the unfolding rates. With increasing repeat number, the unfolding rates decrease moderately in TPR proteins, but enormously for our stable consensus DARPins. This behavior can be rationalized, following the Ising model, by considering the folding process as a nucleation process where the formation of a small assembly of repeats (probably one single consensus repeat in the DARPins) triggers the whole folding cascade. The reverse process (i.e., unfolding), on the other hand, requires the progressive disruption of all the "condensed" folded repeats and is thus dependent on protein length.

Materials and Methods

Design and synthesis of DNA-encoding AR proteins

Oligonucleotides were obtained from Microsynth (Balgach, Switzerland), following the assembly strategy described previously:[18]

INT5 (forward) : 5'-TTCCGCGGATCCTAGGAAG-ACCTGACGTTAACGCT-3'

PRO1 (forward) : 5'-CTGACGTTAACGCTAAAGA-CAAAGACGGTTACACTCCGCT-GCACCTGGC-3'

PRO2 (forward) : 5'-ACTCCGCTGCACCTGGCTG-CTCGTGAAGGTCACCTGG-AAATCG-3'

PRO3 (reverse) : 5'-ACGTCAGCACCAGCCTTC-AGCAGAACTTCAACGATTTC-CAGGTGACC-3'

‡ http://depts.washington.edu/bakerpg/

PRO4 (reverse) : 5'-TTTGGGAAGCTTCTAAGGT-
CTCACGTCAGCACCAG-3'.

The full-consensus AR was generated by assembly PCR using the oligonucleotides PRO1, PRO2, PRO3, PRO4 and INT5, and Vent® polymerase (5 min at 95 °C; followed by 25 cycles of 30 s at 95 °C, 1 min at 50 °C and 30 s at 72 °C; followed by 5 min at 72 °C; standard Vent® polymerase buffer with a final concentration of 3.5 mM MgSO$_4$). The PCR product was cloned via BamHI/HindIII into pPANK,[18] a pQE30 (QIAgen, Germany) derivative lacking the BbsI and BsaI sites, and sequenced using standard techniques. The resulting plasmid was termed pPRO.

Using the plasmids pPRO, pEWT (a pPANK derivative containing the N-terminal capping AR) and pWTC (a pPANK derivative containing the C-terminal capping AR), the DNA encoding the six AR proteins was generated by a ligation procedure using the type IIs restriction enzymes BpiI and BsaI, as described previously.[18]

Protein expression and purification

The repeat proteins NI$_1$C–NI$_6$C were expressed as follows: 50 ml of overnight cultures of *E. coli* XL1-Blue (LB medium, 1% glucose and 100 mg/l ampicillin; 37 °C) was used to inoculate 1-l cultures (LB medium, 1% glucose and 50 mg/l ampicillin; 37 °C). At OD$_{600}$ = 0.7, the cultures were induced with 500 μM IPTG, and incubation was continued for 4 h. The cultures were harvested by centrifugation at 3300g for 10 min at 4 °C, and the resulting pellets were resuspended in 40 ml of 50 mM Tris–HCl (pH 8) and 500 mM NaCl. The cells were lysed using a French press, and the lysate was centrifuged again at 8000g for 15 min at 4 °C, and glycerol (final concentration, 10%) and imidazole (final concentration, 20 mM) were added to the resulting supernatant. The proteins were purified over a Ni-NTA column (column volume, 3.8 ml) in accordance with the manufacturer's instructions (QIAgen). The protein was then dialyzed overnight against 50 mM sodium phosphate buffer (pH 7.4) and 150 mM NaCl (PBS$_{150}$). Purity was checked with 15% SDS-PAGE, the monomeric state was verified by gel filtration combined with multiangle light scattering (miniDAWN, Wyatt, Germany; Astra software) and the correct molecular mass was verified by mass spectrometry.

CD spectroscopy

The CD signal at 222 nm was recorded on a Jasco J-715 instrument (Jasco, Japan) equipped with a computer-controlled water bath, using a cylindrical quartz cell of 1 mm pathlength. To measure denaturant-induced equilibrium unfolding, CD data were collected at 222 nm (measurement intervals, 5 s), 2-nm bandwidth and 4-s response time, and each data point was averaged over 2 min. Thermal unfolding was recorded by continuous heating with a temperature gradient of 0.5 °C min^{-1} from 15 °C to 95 °C. CD data were collected at 222 nm (measurement intervals, 5 s), 2-nm bandwidth and 4-s response time. Reversibility was determined from the recovery of ellipticity after cooling.

All CD experiments were performed in PBS$_{150}$ using 10 μM purified protein, and a baseline correction was made with the buffer. The CD signal was converted to mean residue ellipticity (Θ_{MRW}) using the concentration of the sample determined spectrophotometrically at 280 nm.

Fluorescence spectroscopy

Tyrosine fluorescence was measured by excitation at 274 nm and by recording the emission spectra from 290 nm to 350 nm using a PTI Alpha Scan spectrofluorimeter (Photon Technologies, Inc.). Slid widths of 5 nm were used for both excitation and emission. Samples were prepared as for the CD measurements. After buffer correction, the intensity of the emission maximum at 306 nm was plotted against denaturant concentration.

Equilibrium unfolding

The transitions were monitored using both the CD signal and tyrosine fluorescence. For measuring denaturant-induced equilibrium unfolding curves, the samples were equilibrated overnight at the corresponding GdnHCl concentrations at 20 °C. The GdnHCl concentrations were determined by refractive index.

The data were fitted by assuming both classical cooperative folding models and an Ising model (see below). Where indicated, a two-state model[33] was used according to Eq. (1):

$$S_{obs}(D) = (S_U + m_U[D])f_U + (S_N + m_N[D])f_N \quad (1)$$

where $S_{obs}(D)$ is the observed quantity (e.g., CD signal or fluorescence signal) as a function of denaturant concentration [D], S_U is the signal of the unfolded state (extrapolated to zero [D]), S_N is the signal of the native state at zero [D], and f_U and f_N are unfolded and native fractions, respectively, defined as:

$$f_U = 1 - f_N = K_U/(1 + K_U) \quad (2)$$

where $K_U = [U]/[N]$ is the equilibrium constant of denaturation. Where indicated, a sequential three-state model[34,35] for equilibrium denaturation was used according to Eqs. (3)–(9):

$$U \rightleftharpoons I \rightleftharpoons N \quad (3)$$

with

$$K_{UI} = [I]/[U] \quad (4)$$

$$K_{IN} = [N]/[I] \quad (5)$$

$$f_U = 1/(1 + K_{UI} + K_{UI}K_{IN}) \quad (6)$$

$$f_I = K_{UI}/(1 + K_{UI} + K_{UI}K_{IN}) \quad (7)$$

$$f_N = K_{UI}K_{IN}/(1 + K_{UI} + K_{UI}K_{IN}) \quad (8)$$

$$S_{obs}(D) = (S_U + m_U[D])f_U + (S_I + m_I[D])f_I \\ + (S_N + m_N[D])f_N \quad (9)$$

where U, I and N represent the unfolded, intermediate and native states of the protein, and f and S are the fractional population and spectroscopic signals of the states indicated by the subscripts, respectively. Both $\ln(K_{IN})$ and $\ln(K_{UI})$ are assumed to be linearly dependent on [D]. Data were fitted using ProFit (Quantum Soft, Switzerland).

Kinetic folding and unfolding

Kinetic experiments were performed with a PiStar-180 stopped-flow instrument (Applied Photophysics, UK). The CD signal was monitored at 225 nm, and tyrosine fluorescence changes were monitored with a 295-nm cutoff filter. For NI$_3$C, an optical pathlength of 2 mm was used, and the final concentration was 18 μM. NI$_1$C and NI$_2$C were measured with a 10-mm pathlength, and the final protein concentrations were 4–5 μM. All refolding and unfolding reactions were measured in PBS$_{150}$ at 20 °C. Refolding experiments were performed as follows: the protein NI$_1$C was denatured in 2.5 M, the protein NI$_2$C was denatured in 5.5 M and the protein NI$_3$C was denatured in 6.8 M GdnHCl. Refolding was initiated by rapid mixing of 1 vol of denatured protein solution with 10 vol of buffer containing various concentrations of denaturant. Unfolding rates were measured by rapid mixing of 1 vol of native protein solution with 10 vol of denaturing buffer with different GdnHCl concentrations (>1.4 M for NI$_1$C; >4.1 M for NI$_2$C; >5.6 M for NI$_3$C). Reactions with a half-life longer than about 12 s were mixed manually.

Ten to 25 kinetic traces were averaged for each GdnHCl concentration and fitted to either a single exponential function or a double exponential function according to Eqs. (10) and (11):

$$S_{obs}(t) = a \, e^{-\lambda t} + b \qquad (10)$$

$$S_{obs}(t) = a_1 \, e^{-\lambda_1 t} + a_2 \, e^{-\lambda_2 t} + b \qquad (11)$$

where a_1 or a_2 represents the change in spectroscopic signal of phase 1 or phase 2; λ_1 or λ_2 represents the observed rate constants of phase 1 or phase 2; and b represents the spectroscopic signal after the reaction had reached equilibrium.

The observed rate constants λ_i were analyzed as a function of the GdnHCl concentration, according to a kinetic sequential three-state model with on-pathway intermediate:[28]

$$U \underset{k_{IU}}{\overset{k_{UI}}{\rightleftharpoons}} I \underset{k_{NI}}{\overset{k_{IN}}{\rightleftharpoons}} N \qquad (12)$$

where all four microscopic rate constants k_{UI}, k_{IU}, k_{IN} and k_{NI} are defined by the solution of a quadratic equation according to Eqs. (13)–(15):

$$\lambda_{1,2} = \frac{-B \pm \sqrt{B^2 - 4C}}{2} \qquad (13)$$

with

$$B = -(k_{UI} + k_{IU} + k_{IN} + k_{NI}) \qquad (14)$$

$$C = k_{UI}(k_{IN} + k_{NI}) + k_{IU}k_{NI} \qquad (15)$$

and the refolding and unfolding rates are defined according to Eqs. (16) and (17):

$$k_f^0 = k_{UI}^0 k_{IN}^0 / (k_{IU}^0 + k_{IN}^0) \qquad (16)$$

$$k_u^0 = k_{NI}^0 k_{IU}^0 / (k_{NI}^0 + k_{IU}^0) \qquad (17)$$

For all k_{ij}, the relation $\ln k_{ij} = \ln k_{ij}^0 + m_{ij}[D]$ was assumed, where k_{ij} represents the denaturant-dependent rate constant, k_{ij}^0 is the rate constant in the absence of denaturant and m_{ij} is the denaturant dependence of $\ln k_{ij}$.

For NI$_1$C, a three-state model with a metastable high-energy intermediate was used,[29] as all measured kinetics were monoexponential. For NI$_2$C, the single observed refolding phase and both unfolding phases were simultaneously fitted to Eq. (13) using ProFit. For NI$_3$C, the folding and unfolding rate constants were determined by fitting the denaturant dependencies of the single refolding phase and the slowest unfolding phase, using the relation $\ln k_{ij} = \ln k_{ij}^0 + m_{ij}[D]$.

Ising model for the folding and unfolding of NI$_x$C

As an alternative to cooperative folding models (see above), an Ising-like model was used to study the dependence of the stability of the consensus DARPins NI$_x$C on denaturant concentration. According to this model, each repeat of the protein is considered as an independent folding unit with a free energy of unfolding that depends linearly on the denaturant concentration [D]. Adjacent folded repeats interact by a stabilizing potential, whose magnitude is independent of [D] but requires that both repeats be folded.

The effective free energy of a configuration of the repeat protein ΔG_{conf}, where some of the repeats may be folded and some may not, can be written according to Eq. (18):

$$\Delta G_{conf}(s_i; D) = \Delta G(D) \sum_{i=2}^{L-1} s_i + \Delta G'(D)(s_1 + s_L)$$
$$+ J \sum_{i=2}^{L-1} s_i s_{i+1} \qquad (18)$$

where s_i is a variable that describes the state of the ith repeat ($s_i = 1$ when folded; $s_i = 0$ otherwise), and L is the total number of repeats in the protein. $\Delta G(D) = \Delta G_0 + m[D]$ is the free energy of folding of a consensus repeat that depends linearly on the denaturant concentration. A different stability $\Delta G'(D) = \Delta G_0' + m'[D]$ has been introduced to describe the different characteristics of the C- and N-terminal capping repeats with respect to the consensus repeats.

The interaction parameter J has been chosen as independent of [D] to reduce the number of free parameters of the model. More complex models have been considered, but, in our opinion, the improvement of the fit did not justify the increase in the number of free parameters of the model.

The probability of each state of the proteins $P_L(\{s_i\})$ (i.e., the probability of observing a certain set of folded/unfolded repeats $\{s_i\}$ for a protein with L repeats) under any concentration of denaturant [D] is obtained using the standard canonical formalism (Eq. (19)):

$$P_L(s_i; [D]) = \frac{e^{-\Delta G_{conf}(s_i;[D])/kT}}{\sum_{s_i} e^{-\Delta G_{conf}(s_i;[D])/kT}} \qquad (19)$$

where k is the Boltzmann constant, T is the temperature and the sum in the denominator represents the canonical partition function (extended to all possible states of the protein; i.e., all the possible combinations of folded/unfolded repeats).

Finally, it is possible to compute the average fraction of folded repeats, at each value of [D], according to Eq. (20):

$$f_L([D]) = \sum_{s_i} P_L(s_i; [D]) \frac{1}{L} \sum_{i=1}^{L} s_i \qquad (20)$$

The parameters of the model have been determined by fitting the denaturant-induced equilibrium denaturation data of proteins NI$_1$C, NI$_2$C and NI$_3$C, which were obtained at room temperature and monitored by CD. Equation (21) was used to fit the observed CD signal:

$$S_{obs} = (S_U + m_U[D])(1 - f_L([D])) + (S_N + m_N[D])f_L([D]) \qquad (21)$$

Note that, in Eq. (21), f_L refers to the fraction of folded repeats, while in Eqs. (1) and (9), f_N refers to the fraction of folded whole molecules. The baselines of the CD signal for the folded and unfolded states have been considered as independent of [D] where possible (i.e., to have a slope of zero), in order to keep the number of free parameters as low as possible. In the case of NI_1C and NI_2C, the very long posttransition baseline had to be considered as linearly dependent on [D] because the approximation of a slope of zero was clearly not suitable. Different baselines were allowed for the different protein constructs. The least-squares fit has been obtained using the Levenberg–Marquardt algorithm, where both the parameters of the model and the coefficients of the baselines were allowed to vary.

Acknowledgements

We thank Drs. Ilian Jelesarov and Ben Schuler for fruitful discussions, Dr. V. Sathya Devi for help with the stopped-flow instrument, Dr. Christophe Bodenreider (Biozentrum, Basel, Switzerland) for help with data fitting using ProFit and Enrico Guarnera for his valuable help. G.S. would like to thank Drs. Amedeo Caflisch and Gianluca Interlandi for stimulating discussions. This work was supported by the Swiss National Science Foundation and the National Center of Competence in Research Structural Biology.

Supplementary Data

Supplementary data associated with this article can be found, in the online version, at doi:10.1016/j.jmb.2007.11.046

References

1. Andrade, M. A., Perez-Iratxeta, C. & Ponting, C. P. (2001). Protein repeats: structures, functions, and evolution. *J. Struct. Biol.* 134, 117–131.
2. Bork, P. (1993). Hundreds of ankyrin-like repeats in functionally diverse proteins: mobile modules that cross phyla horizontally? *Proteins: Struct. Funct. Genet.* 17, 363–374.
3. Sedgwick, S. G. & Smerdon, S. J. (1999). The ankyrin repeat: a diversity of interactions on a common structural framework. *Trends Biochem. Sci.* 24, 311–316.
4. Koradi, R., Billeter, M., Wüthrich, K. (1996). MOLMOL: a program for display and analysis of macromolecular structures. *J. Mol. Graphics*, 14, 51–55, 29–32.
5. Kohl, A., Binz, H. K., Forrer, P., Stumpp, M. T., Plückthun, A. & Grütter, M. G. (2003). Designed to be stable: crystal structure of a consensus ankyrin repeat protein. *Proc. Natl Acad. Sci. USA*, 100, 1700–1705.
6. Tang, K. S., Guralnick, B. J., Wang, W. K., Fersht, A. R. & Itzhaki, L. S. (1999). Stability and folding of the tumour suppressor protein p16. *J. Mol. Biol.* 285, 1869–1886.
7. Tang, K. S., Fersht, A. R. & Itzhaki, L. S. (2003). Sequential unfolding of ankyrin repeats in tumor suppressor p16. *Structure (Cambridge)*, 11, 67–73.
8. Interlandi, G., Settanni, G. & Caflisch, A. (2006). Unfolding transition state and intermediates of the tumor suppressor p16INK4a investigated by molecular dynamics simulations. *Proteins: Struct. Funct. Genet.* 64, 178–192.
9. Zeeb, M., Rosner, H., Zeslawski, W., Canet, D., Holak, T. A. & Balbach, J. (2002). Protein folding and stability of human CDK inhibitor p19(INK4d). *J. Mol. Biol.* 315, 447–457.
10. Low, C., Weininger, U., Zeeb, M., Zhang, W., Laue, E. D., Schmid, F. X. & Balbach, J. (2007). Folding mechanism of an ankyrin repeat protein: scaffold and active site formation of human CDK inhibitor p19 (INK4d). *J. Mol. Biol.* 373, 219–231.
11. Mello, C. C. & Barrick, D. (2004). An experimentally determined protein folding energy landscape. *Proc. Natl Acad. Sci. USA*, 101, 14102–14107.
12. Zweifel, M. E. & Barrick, D. (2001). Studies of the ankyrin repeats of the *Drosophila melanogaster* Notch receptor: 2. Solution stability and cooperativity of unfolding. *Biochemistry*, 40, 14357–14367.
13. Mello, C. C., Bradley, C. M., Tripp, K. W. & Barrick, D. (2005). Experimental characterization of the folding kinetics of the notch ankyrin domain. *J. Mol. Biol.* 352, 266–281.
14. Bradley, C. M. & Barrick, D. (2006). The notch ankyrin domain folds via a discrete, centralized pathway. *Structure*, 14, 1303–1312.
15. Mosavi, L. K., Williams, S. & Peng Zy, Z. Y. (2002). Equilibrium folding and stability of myotrophin: a model ankyrin repeat protein. *J. Mol. Biol.* 320, 165–170.
16. Lowe, A. R. & Itzhaki, L. S. (2007). Biophysical characterisation of the small ankyrin repeat protein myotrophin. *J. Mol. Biol.* 365, 1245–1255.
17. Lowe, A. R. & Itzhaki, L. S. (2007). Rational redesign of the folding pathway of a modular protein. *Proc. Natl Acad. Sci. USA*, 104, 2679–2684.
18. Binz, H. K., Stumpp, M. T., Forrer, P., Amstutz, P. & Plückthun, A. (2003). Designing repeat proteins: well-expressed, soluble and stable proteins from combinatorial libraries of consensus ankyrin repeat proteins. *J. Mol. Biol.* 332, 489–503.
19. O'Neil, K. T. & DeGrado, W. F. (1990). A thermodynamic scale for the helix-forming tendencies of the commonly occurring amino acids. *Science*, 250, 646–651.
20. Mosavi, L. K., Minor, D. L., Jr. & Peng, Z. Y. (2002). Consensus-derived structural determinants of the ankyrin repeat motif. *Proc. Natl Acad. Sci. USA*, 99, 16029–16034.
21. Interlandi, G., Wetzel, S.K., Settanni, G., Plückthun, A., Caflisch, A. (2007). Characterization and further stabilization of designed ankyrin repeat proteins by combining molecular dynamics simulations and experiments. *J. Mol. Biol.* in press [2007 Sep 21; Epub ahead of print]. doi:10.1016/j.jmb.2007.09.042.
22. Merz, T., Wetzel, S.K., Firbank, S., Plückthun, A., Grütter, M., Mittl, P.R.E. (2007). Stabilizing ionic interactions in a full consensus ankyrin repeat protein. *J. Mol. Biol.* [in press].
23. Myers, J. K., Pace, C. N. & Scholtz, J. M. (1995). Denaturant m values and heat capacity changes: relation to changes in accessible surface areas of protein unfolding. *Protein Sci.* 4, 2138–2148.
24. Yu, H., Kohl, A., Binz, H. K., Plückthun, A., Grütter, M. G. & van Gunsteren, W. F. (2006). Molecular dynamics study of the stabilities of consensus designed ankyrin repeat proteins. *Proteins: Struct. Funct. Genet.* 65, 285–295.
25. Tripp, K. W. & Barrick, D. (2007). Enhancing the stability and folding rate of a repeat protein through

the addition of consensus repeats. *J. Mol. Biol.* **365**, 1187–1200.
26. Kajander, T., Cortajarena, A. L., Main, E. R., Mochrie, S. G. & Regan, L. (2005). A new folding paradigm for repeat proteins. *J. Am. Chem. Soc.* **127**, 10188–10190.
27. Main, E. R., Stott, K., Jackson, S. E. & Regan, L. (2005). Local and long-range stability in tandemly arrayed tetratricopeptide repeats. *Proc. Natl Acad. Sci. USA*, **102**, 5721–5726.
28. Buchner, J. & Kiefhaber, T. (2005). Protocols—analytical solutions of three-state protein folding models. In *Protein Folding Handbook: Part 1*, vol. 1, pp. 402–406, Wiley-VCH Verlag GmbH and Co. KGaA, Weinheim.
29. Bachmann, A. & Kiefhaber, T. (2001). Apparent two-state tendamistat folding is a sequential process along a defined route. *J. Mol. Biol.* **306**, 375–386.
30. Fersht, A. (1999). *Structure and Mechanism in Protein Science: A Guide to Enzyme Catalysis and Protein Folding*. Freeman, New York.
31. Devi, V. S., Binz, H. K., Stumpp, M. T., Plückthun, A., Bosshard, H. R. & Jelesarov, I. (2004). Folding of a designed simple ankyrin repeat protein. *Protein Sci.* **13**, 2864–2870.
32. Plaxco, K. W., Simons, K. T. & Baker, D. (1998). Contact order, transition state placement and the refolding rates of single domain proteins. *J. Mol. Biol.* **277**, 985–994.
33. Pace, C. N. (1975). The stability of globular proteins. *CRC Crit. Rev. Biochem.* **3**, 1–43.
34. Barrick, D. & Baldwin, R. L. (1993). Three-state analysis of sperm whale apomyoglobin folding. *Biochemistry*, **32**, 3790–3796.
35. Pace, C. N. (1986). Determination and analysis of urea and guanidine hydrochloride denaturation curves. *Methods Enzymol.* **131**, 266–280.

Results II

2.2 Characterization and Further Stabilization of Designed Ankyrin Repeat Proteins by Combining Molecular Dynamics Simulations and Experiments (*II*)

This chapter presents the results of molecular dynamics (MD) simulations with NI_1C, NI_2C, NI_3C and several other DARPins. Analysis of the MD runs suggested that unfolding begins with the denaturation of the C-capping repeat. The MD simulation results inspired us to construct DARPins without the capping repeats. These experiments proved the importance of the capping repeats for solubility as well as support the unfolding mechanism suggested from MD simulations. Furthermore, six new C-capping repeats were designed. The stabilities were assessed measuring the equilibrium unfolding of the new proteins (NI_1C and NI_3C mutants) with substituted C-terminal capping repeats. The high stability of the full consensus repeat was explained on the basis of structural considerations (electrostatic interactions) and compared to the less stable N3C library members.

*Interlandi, G., *Wetzel, S. K., Settanni, G., Plückthun, A. & Caflisch, A. (2007). Characterization and further stabilization of designed ankyrin repeat proteins by combining molecular dynamics simulations and experiments. *J. Mol. Biol. 375, 837-854*
(for the coloured version of the article refer to the publisher weblink
http://dx.doi.org/10.1016/j.jmb.2007.09.042)

* contributed equally

II

Results II

Characterization and Further Stabilization of Designed Ankyrin Repeat Proteins by Combining Molecular Dynamics Simulations and Experiments

Gianluca Interlandi†, Svava K. Wetzel†, Giovanni Settanni, Andreas Plückthun* and Amedeo Caflisch*

Department of Biochemistry, University of Zürich, CH-8057 Zürich, Switzerland

Received 15 January 2007; received in revised form 11 August 2007; accepted 6 September 2007
Available online 21 September 2007

Multiple molecular dynamics simulations with explicit solvent at room temperature and at 400 K were carried out to characterize designed ankyrin repeat (AR) proteins with full-consensus repeats. Using proteins with one to five repeats, the stability of the native structure was found to increase with the number of repeats. The C-terminal capping repeat, originating from the natural guanine-adenine-binding protein, was observed to denature first in almost all high-temperature simulations. Notably, a stable intermediate is found in experimental equilibrium unfolding studies of one of the simulated consensus proteins. On the basis of simulation results, this intermediate is interpreted to represent a conformation with a denatured C-terminal repeat. To validate this interpretation, constructs without C-terminal capping repeat were prepared and did not show this intermediate in equilibrium unfolding experiments. Conversely, the capping repeats were found to be essential for efficient folding in the cell and for avoiding aggregation, presumably because of their highly charged surface. To design a capping repeat conferring similar solubility properties yet even higher stability, eight point mutations adapting the C-cap to the consensus AR and adding a three-residue extension at the C-terminus were predicted *in silico* and validated experimentally. The *in vitro* full-consensus proteins were also compared with a previously published designed AR protein, E3_5, whose internal repeats show 80% identity in primary sequence. A detailed analysis of the simulations suggests that networks of salt bridges between β-hairpins, as well as additional interrepeat hydrogen bonds, contribute to the extraordinary stability of the full consensus.

© 2007 Elsevier Ltd. All rights reserved.

Edited by F. Schmid

Keywords: protein denaturation; protein engineering; network of salt bridges; folding pathways; ankyrin repeat proteins

Introduction

The hallmark of repeat proteins is their modular native-state architecture, which has been discovered in a variety of polypeptide families in the last decade.[1–3] The ankyrin repeat (AR) consists of 33 amino acids forming a loop, a β-turn and two antiparallel α-helices connected by a tight turn.[1] Multiple ARs are stacked in a linear array to form a rigid solenoidal native structure, which is stabilized predominantly by interactions between residues that are close in sequence (Fig. 1). Furthermore, the hydrophobic core of repeat proteins has a toroidal shape, unlike globular proteins. AR-

*Corresponding authors. E-mail addresses: plueckthun@bioc.uzh.ch; caflisch@bioc.uzh.ch.
†G.I. and S.K.W. contributed equally to this work.
Present addresses: G. Interlandi, Department of Bioengineering, University of Washington, Seattle, WA, USA; G. Settanni, MRC Center for Protein Engineering, University of Cambridge, Cambridge, UK.
Abbreviations used: AR, ankyrin repeat; DARPins, designed AR proteins; MD, molecular dynamics; PDB, Protein Data Bank; GdnHCl, guanidine hydrochloride; CD, circular dichroism; MALS, multiangle light scattering.

Results II

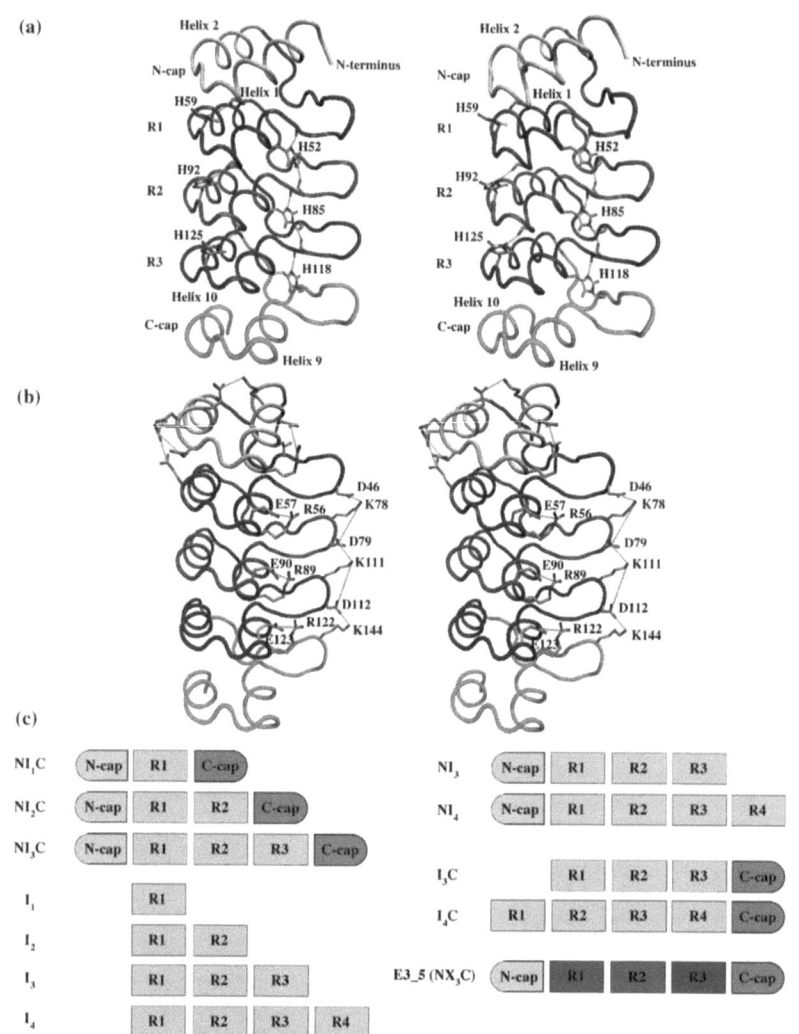

Fig. 1. (a and b) Stereo view of the DARPin NI$_3$C. The structure was modeled from the X-ray structure of E3_5,[12] which is identical in sequence in the N- and C-terminal capping repeats and is 80% identical in the internal repeats. The N-terminus is on top, and individual repeats are emphasized with different colors. Native hydrogen bonds (a) and salt bridges (b) are displayed as red dashed lines. The salt bridges involving the Lys78 and Lys111 side chains on the central loops are conserved in <50% of the simulation frames and are displayed in blue (see NI$_3$C *versus* E3_5). All side chains involved in contacts represented in (a) and (b) are shown in sticks and colored by atom type. (c) Cartoon representation of DARPins. Images (a) and (b) were prepared with the Visual Molecular Dynamics software.[40]

containing proteins are very common in nature and prevalently mediate specific protein–protein interactions.[4]

AR proteins consisting of several identical repeats have been designed and characterized biophysically.[5–7] By combining sequence and structure

Results II

consensus analyses, an AR module was designed with seven randomized positions in the loop and in the first helix.[7,8] Different numbers of this module could be joined to generate combinatorial libraries of AR proteins. Because of the self-complementarity of consensus repeats, the size of the binding site can be altered simply by adding or removing repeats. To reduce the solvent exposure of hydrophobic surfaces, internal modules were flanked by N- and C-terminal "capping" repeats, which were borrowed from the guanine-adenine (GA)-binding protein[9] and slightly modified to approach the consensus and for cloning purposes.[7] Library members were then selected to function as specific binders or even enzyme inhibitors.[7,10,11] Designed AR proteins (DARPins) were shown experimentally to be thermodynamically stable, soluble and highly expressed in native form in bacteria.[7,12]

Up to now, only a few studies have been carried out to gain insight into the folding or unfolding pathways of AR proteins. The unfolding of the four-repeat tumor suppressor p16^{INK4a}, an inhibitor of a cyclin-dependent kinase, starts at the two N-terminal repeats, as suggested by mutagenesis experiments[13] and as verified by molecular dynamics (MD) simulations.[14] These repeats deviate more from the consensus sequence and may thus be intrinsically less stable. The role of topology in the energy landscape of AR proteins has recently been investigated by a simplified structure-based model.[15] The equilibrium folding behavior[16,17] and a kinetic on-pathway intermediate[18] have been experimentally characterized for the *Drosophila* Notch receptor and variants with different numbers of repeats. However, a much more detailed understanding of the folding and unfolding mechanism is essential to shed light on the stabilizing factors of AR proteins, especially in view of the increasing interest in their application in biotechnology.

Here, stabilizing interactions at room temperature and the unfolding pathways of several AR proteins and mutants thereof have been investigated by a combination of equilibrium unfolding experiments and multiple MD runs in explicit water for a total simulation time of >2 μs (Table 1). We first addressed the question of how the number of repeats affects stability. For this purpose, a full-consensus sequence with identical repeats was chosen.[19] We denote these proteins as NI$_x$C, where N and C refer to the N- and C-terminal capping repeats, respectively, I refers to the internal "full"-consensus repeat and the subscript x gives the number of identical internal-consensus repeats. Furthermore, the protein E3_5, a member of the NX$_3$C library[12] (where X denotes a library repeat module) with an 80% sequence identity to the full-consensus repeats, was chosen for MD analysis.

As mentioned above, the primary structure of flanking repeats differs from the consensus design. Hence, interrepeat interfaces involving the terminal repeats are different from interfaces between internal repeats.

To understand the role of the capping repeats in favoring solubility but potentially limiting stability, they were removed both experimentally and in MD simulations at room temperature and at high temperature. Moreover, mutations to further improve the stability of the C-terminal cap were suggested, and the MD simulations and equilibrium unfolding experiments of six NI$_1$C mutants and two NI$_3$C mutants were performed at room temperature.

The aim of the present study, combining simulations and experimental work, is to dissect the architecture of ARs to identify mutations that are critical for stability and to shed light on the role of the capping repeats and the relationship between stability and number of repeats. Furthermore, by an analysis of the high-temperature unfolding mechanism at the atomic level of detail, weak links may

Table 1. Simulation systems

Protein structure	Number of repeats[a]	Number of amino acids	Net charge[b] (electron units)	Box size[c] (Å)	Simulation time[d] (ns)	
					300 K	400 K
NI$_1$C	3 (1)	90	−8	65.1	50, 40[e]	150, 200[e]
NI$_2$C	4 (2)	123	−10	80.6	50	100, 150[e]
NI$_3$C	5 (3)	156	−12	80.6/99.2	50	100
E3_5	5 (3)	156	−16	80.6/99.2	50	100
I$_1$	1 (1)	22[f]	−1	49.6	40	40
I$_2$	2 (2)	55[f]	−3	55.8	50	150
I$_3$	3 (3)	88[f]	−5	62.0	50	150

[a] Total number of repeats and, in parentheses, the number of noncapping repeats. In NI$_1$C, NI$_2$C and NI$_3$C, the N- and C-capping repeats flank the indicated number of identical consensus repeats. In I$_1$, I$_2$ and I$_3$, these capping repeats are missing, and only the indicated number of identical repeats is present. E3_5 is a member of the NX$_3$C library and differs at about seven positions per internal repeat, but has identical capping sequences.
[b] Total net charge of the protein at pH 7.
[c] Initial side length of the cubic box. The box adjusts its volume according to the given temperature and pressure during the simulation (see Materials and Methods). A larger water box is used to simulate NI$_3$C and E3_5 at 400 K.
[d] The total simulation time of the 300-K runs includes the initial 10 ns that were discarded during the analysis (see Materials and Methods).
[e] Two 300-K trajectories of NI$_1$C and two 400-K trajectories of NI$_1$C and NI$_2$C were run with different initial random assignments of the velocities.
[f] The 11 residues preceding the first helix of the first repeat were deleted because large displacements were observed in preliminary simulations of these proteins.

become apparent, in turn helping to understand the experimental unfolding data and the further design of ankyrins.

Results

Fluctuations and stabilizing interactions at room temperature

Comparison with crystallographic B-factors

In the loop regions of E3_5, larger fluctuation values of C_α atoms were observed along the MD trajectory at 300 K than those derived from B-factors (Fig. 2a). This discrepancy is probably due to intermolecular contacts between Loops 1 and 2 and two neighboring protein molecules in the crystal [Protein Data Bank (PDB) code 1MJ0[12]]. On the other hand, very low B-factors and fluctuations during the MD simulations characterize the helical and tight-turn regions.

Stabilizing interactions

DARPins are stable at 300 K. The values of the C_α root mean square deviation (RMSD) averaged over the interval 10–50 ns are 1.55±0.32 Å, 1.94± 0.24 Å, 1.66±0.30 Å and 1.87±0.30 Å for NI_1C, NI_2C, NI_3C and E3_5, respectively. The corresponding all-atom RMSDs are 2.43±0.20 Å, 2.70± 0.21 Å, 2.38±0.24 Å and 2.59±0.24 Å. Several polar interactions observed during 300-K runs contribute to the observed stability (see Materials and Methods for the definition of native contacts from

the simulations). Interesting examples are the His59, His92 and His125 side chains (numbering according to PDB file 1MJ0) at the tight turn between helices 1 and 2 of each repeat (Fig. 1a), which are involved as donors in hydrogen bonds with carbonyl groups in the C-terminal turn of the first helix of the respective preceding repeat. These histidine side chains not only provide an inter-repeat interaction but also contribute to the shielding of the interrepeat interface from the solvent. Additional hydrogen bonds involve the side chains of His52, His85 and His118, which are part of a conserved TPLH motif at the beginning of the first helix of each internal repeat, and the backbone carbonyl group of Ala75, Tyr81, Ala108, Tyr114 and Ala141 in the loop of the following repeat (Fig. 1a). Furthermore, each of these histidines accepts a hydrogen bond from the main-chain NH of the residue $n-3$ and thereby links two adjacent repeats. Shielding of these hydrogen bonds from the solvent is provided mainly by the side chain of the tyrosine in the loops of NI_3C (positions 48, 81 and 114).

NI_3C versus E3_5

In the 300-K simulations, there is a slightly higher number of hydrogen bonds in the full-consensus protein NI_3C than in the library member E3_5, in particular in the internal repeats (Table 2), which is consistent with the smaller fluctuations in NI_3C than in E3_5 (Fig. 2a). Furthermore, the larger number of charged residues in NI_3C compared to E3_5 (48 *versus* 36) leads to an increased number of salt bridges in NI_3C (Table 2 and Fig. 1b). In each

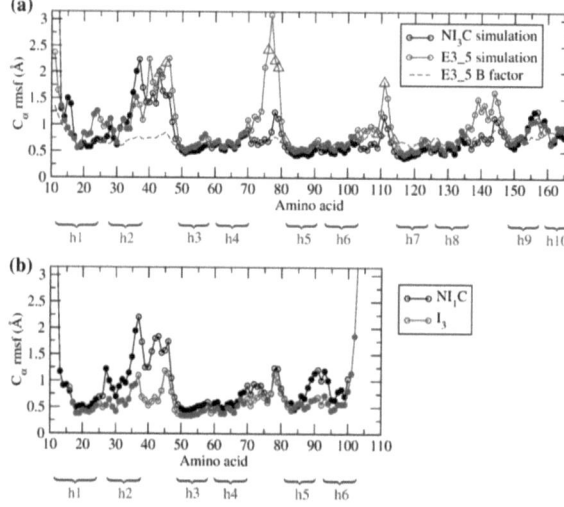

Fig. 2. C_α root mean square fluctuations (RMSFs) at 300 K. (a) NI_3C *versus* E3_5. Values of RMSF derived from crystallographic B-factors of the E3_5 C_α atoms (PDB file 1MJ0) were calculated using the formula $RMSF_{i,\exp} = \sqrt{\frac{3}{8\pi^2}B_i}$, where B_i is the B-factor of C_α residue i. The interval 10–50 ns of each trajectory was used to calculate RMSF values. For each trajectory, a very similar behavior is observed for the interval 10–50 ns and for the eight 5-ns intervals (not shown). Blue triangles indicate the residues where E3_5 differs from NI_3C in primary sequence. (b) NI_1C *versus* I_3 (50-ns run). In the 40-ns run, NI_1C has a C_α RMSF sequence profile similar to that in the 50-ns trajectory. The amino acids located in the helices are represented as filled circles. Helical segments are emphasized by curled braces below the x-axis. h_1 to h_{10} denote helices 1–10.

Results II

Table 2. Native hydrogen bonds, salt bridges and C_α contacts

Protein structure	Total	N	R1	R2	R3	C	N-R1[a]	R1-R2	R2-R3	Rx-C[b]
Hydrogen bonds intraprotein										
X-ray[c]	147	25	26	27	25	25	3	3	6	7
E3_5	113	20	20	20	21	21	1	2	4	4
NI$_3$C	122	19	22	24	22	21	2	4	5	3
NI$_2$C	100	21	21	26		24	2	4		2
NI$_1$C	63	18	19			21	2			3
I$_3$	70		16	26	20			4	4	
I$_2$	37		17	17				3		
Salt bridges										
X-ray[c]	3	1	1	0	0	1	0	0	0	0
E3_5	3	2	0	0	0	0	1	0	0	0
NI$_3$C	9	4	1	1	1	0	1	0	0	1
NI$_2$C	7	2	1	1		1	1	0		1
NI$_1$C	5	3	1			0	1			0
I$_3$	3		1	1	1			0	0	
I$_2$	0		0	0				0		
C_α contacts										
X-ray[c]	290	44	43	43	42	39	20	19	21	19
E3_5	250	43	39	37	41	34	17	9	16	14
NI$_3$C	258	41	39	43	43	37	14	13	15	13
NI$_2$C	207	43	38	43		40	16	15		12
NI$_1$C	148	42	42			36	14			14
I$_3$	149		34	43	45			12	15	
I$_2$	87		34	44				9		

The values listed refer to polar interactions or C_α contacts present in >50% of the simulation time at 300 K (see Materials and Methods for distance threshold definitions). N and C denote the N- and C-caps, respectively, whereas R1 to R3 denote noncapping repeats 1 to 3, respectively (Fig. 1a).
[a] Interactions between the N-terminal cap and first repeat; the other interactions are denoted analogously.
[b] $x=3$ for E3_5 and NI$_3$C; $x=2$ for NI$_2$C; and $x=1$ for NI$_1$C.
[c] Crystallographic structure of E3_5 (PDB 1MJ0).

internal repeat of NI$_3$C, there is a salt bridge between an arginine and a glutamate, which are nearest neighbors in sequence and located at the C-terminal turn of the first helix. The Lys78 and Lys111 side chains (in the loop of repeats R2 and R3, respectively) are involved in a salt bridge with either the neighboring Asp79 (43% of the time) and Asp112 (23% of the time), respectively, or Asp46 (6% of the time) and Asp79 (39% of the time), respectively, located in the loop of the preceding repeat (Fig. 1b and Supplementary Fig. 2). Moreover, in NI$_3$C, the side chains of Asp112 and Lys144 are always at salt-bridge distance. These salt-bridge networks are likely to contribute to the higher thermodynamic stability[20,21] of NI$_3$C than E3_5 observed in the equilibrium unfolding experiments reported below.

I_1–I_3

To test the influence of the caps, structures were generated *in silico* where the capping repeats have been removed. The proteins I$_2$ and I$_3$, consisting only of two and three identical repeats, respectively, were stable at room temperature during the total simulation time of 50 ns. In contrast, I$_1$, consisting of one single AR, reached a C_α RMSD from the starting conformation of 6 Å already after about 30 ns. These simulation results report only on the kinetic stability of the folded state, which is the height of the activation barrier towards unfolding. Yet, they are consistent with experimental data indicating that at least two repeats are necessary for thermodynamic stability[22] (i.e., the energy difference between the

folded state and the unfolded state). Interestingly, there are more hydrogen bonds in I$_3$ than in NI$_1$C (Table 2), which is likely to be one of the reasons for the smaller fluctuations in I$_3$ than in NI$_1$C (Fig. 2b).

In particular, NI$_1$C and all proteins with capping repeats lack two interrepeat hydrogen bonds involving the N-capping repeat, which are present between the internal repeats of NI$_x$C. These hydrogen bonds involve the side chain of the histidine of the TPLH motif, located at the beginning of helix 1 in the consensus design (see Stabilizing Interactions and Fig. 1a). The N-capping repeat contains, at the corresponding position (which is residue 19 according to the numbering in the PDB file 1MJ0), a leucine instead of a histidine.

Similarly, the C-cap contains, at the interhelical tight turn, an asparagine instead of a histidine (e.g., position 158 in NI$_3$C between helices 9 and 10). For this reason, the C-cap lacks the interrepeat hydrogen bond that is present between the designed consensus repeats stabilizing the macrodipole of the first helix of the preceding repeat (see Stabilizing Interactions and Fig. 1a). This observation was taken into account when suggesting the Asn92His mutation in NI$_1$C (see Mutations in the C-Terminal Cap). Furthermore, the terminal helix of the C-cap is three amino acids shorter than the full-consensus sequence. The missing amino acids are lysine, alanine and glycine. This causes the repeat adjacent to the C-cap in NI$_3$C to present a larger solvent-exposed hydrophobic surface (167 ± 21 Å^2) than the central repeat (140 ± 19 Å^2) in the 300-K run. Moreover, the alanine present at the

C-terminal end of the full-consensus repeats increases the helical propensity of the C-terminal helix. This evidence was considered when suggesting a Lys-Ala-Ala extension of the C-cap terminal helix in NI_1C (see below).

Structural stability and high-temperature unfolding mechanism

Correlation between structural stability and number of repeats

The 400-K simulations with the NI_xC and I_x proteins (x = 1, 2 and 3) allow us to analyze the influence of the number of repeats on structural stability (i.e., the kinetic stability of the native state). As mentioned above, I_1 is not stable at 300 K and fully unfolds after 15 ns at 400 K. At 400 K, >75% of the native interrepeat C_α contacts are lost at about 20 ns and 60 ns in I_2 and I_3, respectively (Fig. 3). Using the same criterion (i.e., loss of >75% of the native interrepeat C_α contacts), full unfolding is observed at about 40 ns and 160 ns in the NI_1C runs and at 90 ns in the 100-ns run of NI_2C, whereas complete unfolding is not reached during the 150-ns run of NI_2C and for NI_3C. Despite the very limited statistics that do not allow the evaluation of unfolding rates, the simulation results are consistent with the experimental observation that stability increases with the number of repeats. In fact, the experimentally measured thermodynamic stability has been shown to correlate with the number of repeats and, furthermore, to be due to slower rates of unfolding rather than faster folding rates for NI_xC (x = 1, 2, ..., 6).[19] Similar results have been found for a series of tetratricopeptide repeat proteins,[23] and consistent equilibrium data have been reported for the deletion and duplication of repeats of the Notch receptor ankyrin domain.[16,17] However, this independence of folding rate from repeat number might not always be the case when consensus ARs are mixed with naturally occurring ARs. For example, the insertion of one to two consensus ARs into the five N-terminal repeats of Notch causes an increase in the folding rate.[24]

Sequence of events during unfolding at 400 K

The C-terminal repeat unfolds first for the AR proteins E3_5 and for all NI_xC molecules (Fig. 4 and Supplementary Figs. 3 and 4). The only exception is the 150-ns run of NI_2C, where the helical content and about one-third of the interrepeat contacts of the C-terminal repeat (Fig. 3 and Supplementary Figs. 3 and 4) are partially conserved with an average value of the C_α RMSD from the initial conformation of 6.9±0.5 Å during the last 10 ns. During the unfolding of the C-terminal repeat in all other molecules, the other repeats remain almost completely folded (Supplementary Fig. 4) and conserve most of their native interrepeat C_α contacts (Supplementary Fig. 4e). The C-terminal cap dislocates as a

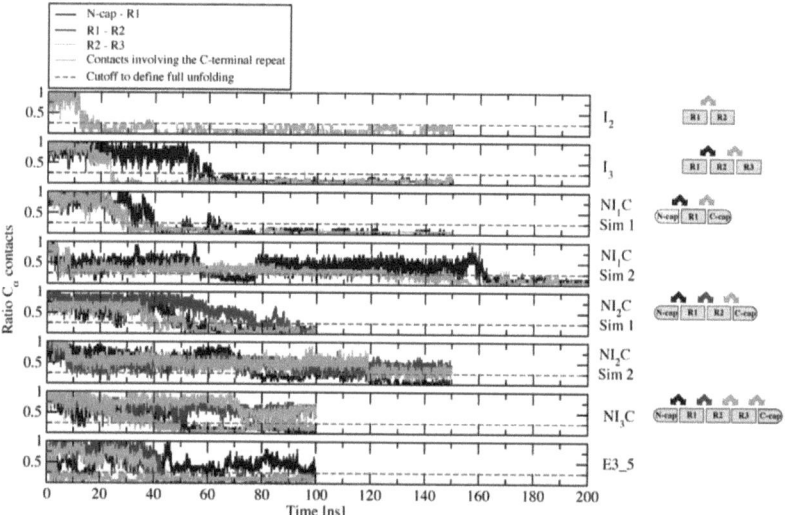

Fig. 3. Time series of the percentage of conserved native interrepeat C_α contacts at 400 K. A cartoon representation of the corresponding DARPin is presented on the right of the plot, with colored arrows indicating contacts between certain repeats. The colors correspond to the curves describing the time course of the respective interactions. For both NI_1C and NI_2C, "Sim 1" and "Sim 2" denote shorter and longer runs, respectively.

Results II

Ankyrin Repeat Protein Unfolding

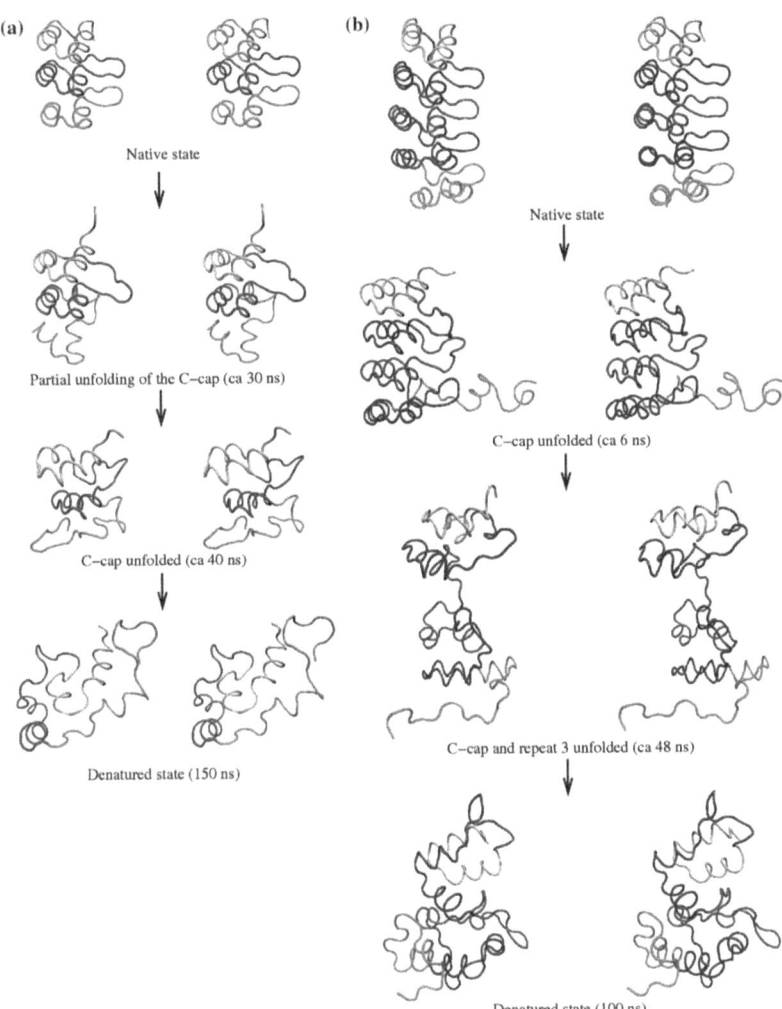

Fig. 4. Stereo view of representative conformations from the 400-K simulations of (a) NI_1C and (b) E3_5.

mostly intact unit first: The rupture of most interrepeat native C_α contacts involving the C-terminal cap (Supplementary Fig. 4e) precedes the unfolding of the helices making up the C-terminal cap (Supplementary Fig. 4c and d). The observation that the C-cap unfolds before all other repeats in the MD simulations may explain the first transition (before the main transition) in the experimentally determined guanidine hydrochloride (GdnHCl)-induced denaturation curve of NI_3C (see below).

After 100 ns of simulation with NI_3C, only the C-terminal cap is unfolded (Supplementary Figs. 3 and 4); thus, it is not possible to extract the complete sequence of unfolding events from the NI_3C run. In the other proteins, most of the interrepeat tertiary contacts are lost within the time scale of simulation (Fig. 3). The sequence of unfolding events is as follows (nomenclature as in Fig. 1; see Supplementary Figs. 3 and 4): C-cap/R1/N-cap in both runs of NI_1C (see also Fig. 4a), C-cap/N-cap/R1/R2 in the

Results II

100-ns run of NI_2C (as mentioned above, in the 150-ns run of NI_2C, the helical content was partially conserved) and C-cap/R2/R3/R1/N-cap in E3_5 (see also Fig. 4b). The unfolding of the central repeat R2 of E3_5 is preceded by the rupture of the hydrophobic interface between repeats R1 and R2, as indicated by the decrease of their native interrepeat C_α contacts (turquoise line in Fig. 3). The fact that the unfolding of repeat R2 of E3_5 directly follows the denaturation of the C-cap is consistent with the smaller number of native interrepeat hydrogen bonds between repeats R1 and R2 of E3_5, compared to NI_3C (Table 2), and to the relatively small number of native C_α contacts between repeats R1 and R2 of E3_5 (Table 2). Moreover, the fact that some internal repeats (e.g., repeat R2 of E3_5 and repeat R1 of NI_1C) unfold before the N-terminal cap provides evidence that the very high temperature (400 K) does not lead to artificial deformations at the protein surface in the simulations. Thus, it can be excluded that the observed early unfolding of the C-cap in the simulations is an artifact caused by the high temperature. Interestingly, the N-terminal cap seems to be more stable than the C-terminal cap, which is consistent with the fact that the N-terminal cap is more similar to the consensus repeat.

Denatured state

During the interval 155–200 ns in one of the two unfolding runs of NI_1C, helix 2 in the N-capping repeat elongates up to residue 47, with π-helical turns at its C-terminal region (Supplementary Fig. 3). The same elongation of helix 2 in the N-capping repeat takes place for NI_3C, where otherwise only the C-terminal capping repeat unfolded during a total simulation time of 100 ns (Supplementary Fig. 3). Nonnative π-helical structure is also present at residues 126–146 of E3_5 (Supplementary Fig. 3). Similarly, at the end of one of the two unfolding runs of NI_2C, residues 107–121 form a nonnative α-helical structure.

Experimental studies on the equilibrium unfolding of NI_3C and variants without capping repeats

Previously, the stability of several DARPins has been measured by both GdnHCl and thermal denaturation.[7,12] These proteins were unselected members of the DARPin library. They were all highly stable, even though some differences between individual library members can be noted. There was a general trend towards higher stability with an increasing number of repeats. However, since the individual library members of the same length covered a range of stabilities,[25] a quantitative relationship could not be established. Most of the proteins tested have previously shown highly cooperative reversible transitions that were consistent with a two-state equilibrium system.

The full-consensus proteins are even more stable, and the details of the dependence of folding and unfolding rates on the number of repeats in the protein are reported in the accompanying manuscript.[19] By the design of the consensus sequence, the previously variable library positions were now chosen according to the most frequent residues, which turn out to be charged or polar; a possible reason for the stability is that additional favorable electrostatic interactions are formed (see NI_3C versus E3_5).

When the full-consensus protein NI_3C is compared to E3_5, a member of the NX_3C library, two differences in GdnHCl equilibrium denaturation experiments are apparent (Fig. 5). First, the main transition is shifted to higher GdnHCl (by 0.8 M). Second, the denaturation is no more fully cooperative and is not consistent with a two-state equilibrium system. Instead, there is a small "pre-transition" visible at 3.7 M GdnHCl, before the main transition, which occurs at about 5.6 M GdnHCl. The two transitions can be well described by a sequential three-state model, and the calculated ΔG (19.7 kcal/mol) is about double that obtained for E3_5 (11.2 kcal/mol; Fig. 5a).

A possible explanation is that the higher stability of the central domains in NI_3C uncouples the unfolding of one or both of the capping repeats, which may therefore unfold already at a lower denaturant concentration and give rise to an equilibrium intermediate with one or both of the caps detached from the central repeats. At this intermediate state, a portion of the circular dichroism (CD) signal is lost. This explanation is also supported by the unfolding behavior of all NI_xC proteins and E3_5 in the MD simulations where the C-cap unfolds prior to the other repeats.

To test this hypothesis, additional proteins, which were devoid of one or both of the capping repeats, were constructed. We denote them NI_3 or NI_4, to indicate that they have only the N-cap and three or four full-consensus repeats, respectively, and I_3C and I_4C, to indicate that they carry only the C-cap and the number of consensus repeats indicated by the number. Finally, we also created the molecules without any caps, which consist only of the consensus repeats and are named I_3 and I_4 (see Materials and Methods and Supplementary Fig. 1 for the definition of the respective sequences).

It is immediately apparent that the molecules lacking both N- and C-terminal caps show significant amounts of insoluble protein upon expression in Escherichia coli (Fig. 6), in contrast to E3_5 and NI_3C, which are completely soluble. The molecules lacking only one of the capping repeats could be purified from the soluble fraction and were further analyzed using multiangle light scattering (MALS). A portion of NI_3 and NI_4 precipitated after elution from the Ni–NTA column, while I_3C and I_4C remained soluble. MALS analysis showed that the proteins I_3C and I_4C form soluble aggregates, however (data not shown). In contrast, the soluble portion of the proteins NI_3 and NI_4, lacking the C-terminal cap, remains mainly monomeric. However, these proteins do remain aggregation-prone, as they aggregate at intermediate concentrations of

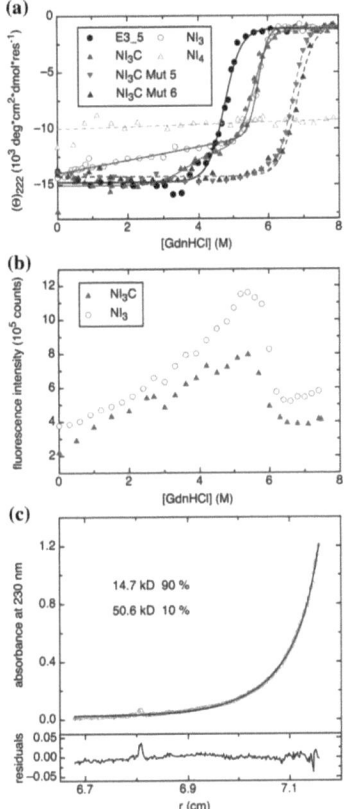

Fig. 5. GdnHCl-induced equilibrium unfolding of ankyrin proteins E3_5 (■), NI₃C (▲), NI₃ (○), NI₄ (△), NI₃C Mut 5 (▼) and NI₃C Mut 6 (▲) at 20 °C followed by (a) CD spectroscopy and (b) tyrosine fluorescence (see Materials and Methods). The lines represent the least-squares fit to the two-state model (for E3_5, NI₃) with a midpoint of denaturation $D_m = 4.8$ M ($\Delta G = 11.2 \pm 0.8$ kcal/mol) for E3_5 and $D_m = 5.7$ M ($\Delta G = 23.6 \pm 2.4$ kcal/mol) for NI₃, or a sequential three-state model[31] with $D_{m1} = 3.7$ M and $D_{m2} = 5.6$ M ($\Delta G = 19.7 \pm 4.6$ kcal/mol) for NI₃C. The protein concentrations were 10 μM (E3_5, NI₃C, NI₃C Mut 5, NI₃C Mut 6), 5 μM (NI₃) and 7 μM (NI₄). The midpoints are calculated from the fit according to $D_{mx} = \Delta G_x^0/m_x$, where x refers to the first or second transition. (c) The oligomerization state of NI₃ was analyzed by analytical ultracentrifugation. NI₃ at 7 μM in 50 mM phosphate, 150 mM NaCl and 2 M GdnHCl was analyzed by sedimentation equilibrium at 35,000 rpm and 20 °C. A global species analysis fit yielded 90% monomeric (14.7 kDa) protein and only 10% higher-molecular-weight species (50.6 kDa).

GdnHCl with increasing protein concentration, and this becomes detectable above a protein concentration of about 7 μM (data not shown). To test our interpretation of the pretransition in NI₃C as being due to the unfolding of the C-cap, we measured the equilibrium unfolding of NI₃ at protein concentrations of 5 μM and 15 μM. The transition point for the curve at 15 μM was shifted to a higher GdnHCl concentration (data not shown), while the main transition of NI₃ at 5 μM was superimposable with that of NI₃C (Fig. 5a). In order to test whether NI₃ is really monomeric at a GdnHCl concentration below its transition, sedimentation equilibrium experiments with two different concentrations of NI₃ and NI₃C at 2 M GdnHCl were performed in an analytical ultracentrifuge. While at 21 μM NI₃ forms significant proportions of higher-molecular-weight species, at 7 μM, the sample of NI₃ consists of 90% monomeric protein (Fig. 5c), as does the sample of NI₃C at a protein concentration of 10 μM (data not shown). The unfolding transition of NI₃ at 5 μM monitored by CD and fluorescence was therefore assigned to that of monomeric protein.

The CD equilibrium unfolding of NI₃ is cooperative and has a similar transition midpoint as NI₃C, which contains the C-terminal cap (Fig. 5a). Importantly, no pretransition at 3.7 M GdnHCl is detected. The absence of this pretransition in NI₃ is indeed consistent with the C-cap being denatured in the equilibrium intermediate of NI₃C. The main transition (and the only transition present for NI₃) is thus interpreted as the cooperative denaturation of the consensus repeats including the N-cap, but not of the C-cap. If the C-cap of NI₃C is selectively denatured in the intermediate at 3.7 M GdnHCl, a molecule identical with NI₃ with a denatured appendage would be obtained, which would be expected to denature under similar conditions as NI₃. Indeed, the equilibrium unfolding curves of NI₃ and NI₃C measured by fluorescence reveal similar transition points at around 5.6 M GdnHCl. This supports the assumption that both monitor the unfolding of monomeric protein. Nevertheless, the nature of the

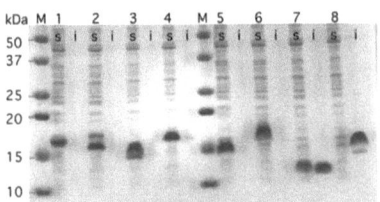

Fig. 6. Expression of ankyrin proteins E3_5 (lane 1, 17.7 kDa) and NI₃C (lane 2, 17.9 kDa), and the cap-lacking constructs NI₃ (lane 3, 14.7 kDa), NI₄ (lane 4, 18.8 kDa), I₃C (lane 5, 14.1 kDa), I₄C (lane 6, 17.6 kDa), I₃ (lane 7, 10.8 kDa) and I₄ (lane 8, 14.3 kDa). At OD₆₀₀ = 0.7, *E. coli* cultures were induced with 0.5 M IPTG and grown for 4 h at 37 °C. After cell lysis using a French press, the proteins in the soluble and insoluble fractions, s and i, were separately analyzed.

Results II

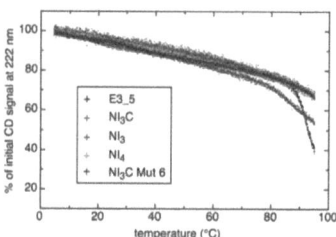

Fig. 7. Thermal melting of E3_5 (black dots), NI$_3$C (red dots), NI$_3$ (blue dots), NI$_4$ (green dots) and NI$_3$C Mut 6 (dark green dots) from 5 °C to 95 °C was followed by CD spectroscopy (see Materials and Methods). The heating gradient was 0.5 °C/min, and melting was only partially reversible (i.e., only 70% of the signal was regained upon cooling).

rather steep slope and of a possible kink in the pretransition baselines remains unclear (Fig. 5b).

The fact that there appears to be only a single cooperative transition in NI$_3$ in equilibrium unfolding measurements does not, of course, preclude that the kinetic process of unfolding is stepwise, with a preferred or random order of repeats unfolding. Indeed, possible unfolding pathways of DARPins emerge from the analysis of the MD simulations.

NI$_4$, which also has no C-cap but one consensus repeat more (and can be thought of as NI$_3$C, with the C-cap being "replaced" by another consensus repeat), shows no transition with GdnHCl, indicating an improved stability over NI$_3$C (Fig. 5a). This indicates that the consensus repeat provides much more stability than the C-capping repeat, albeit at the price of reduced solubility, especially under the conditions of folding in the cell. The natural evolution of the C-capping repeat must therefore have been governed predominantly by solubility and resistance to aggregation as the driving force. Nevertheless, these findings do lead to the question of whether a C-cap that combines high solubility and still shows further improved stability compared to the natural C-cap can be designed.

In thermal denaturation, all proteins (NI$_3$C, NI$_3$ and NI$_4$) are very stable and cannot be melted by heating up to 95 °C (Fig. 7), while E3_5, an unselected library member of the NX$_3$C library, begins denaturation at about 90 °C.

Mutations in the C-terminal cap

The flanking repeats of DARPins are necessary to provide solubility, particularly to allow folding in the cell (see above). However, they have been derived from the naturally occurring GA-binding protein, and their amino acid sequence differs from that of the designed consensus. For this reason, the interface between the capping repeats and their neighboring repeats is not as optimized as the interface between internal repeats. Figure 2b shows that, at 300 K, the C$_\alpha$ atoms of I$_3$ (three full-consensus repeats) fluctuate less than the C$_\alpha$ atoms of NI$_1$C, in particular in the external capping repeats. This is consistent with a better packing of the hydrophobic interface in I$_3$ than in NI$_1$C. Hence, to improve stability further, the internal surface of the C-terminal capping repeat should be engineered using the designed consensus sequence as a guide,[7] while the solvent-exposed residues should remain unmodified, as they are necessary to avoid aggregation.

Six multiple point mutants of NI$_1$C are listed in Table 3, and the side chains involved are shown in Fig. 8a. They are denoted NI$_1$C Mut 1 to NI$_1$C Mut 6. To validate our suggestion *in silico*, two additional 300-K runs were performed for each of the six mutants of NI$_1$C. The point mutants were inspired by bringing the C-terminal repeat closer to the consensus. Ala83Pro introduces a proline present in the consensus repeat, being part of the conserved Thr-Pro-Leu-His motif that is missing in the C-terminal repeat. The mutations Ile86Leu, Ser87Ala, Leu95Ile and Ile[98]Val might improve the packing of the interrepeat hydrophobic interface. Moreover, Ser87Ala could potentially increase the helical propensity. To facilitate the formation of a salt bridge observed in the internal repeats, the mutations Asp89Arg and Asn90Glu were also tested. In addition, Asn92 was changed to a histidine to favor a hydrogen bond with the carbonyl oxygen of residue 56 in the C-terminal turn of the first helix of R1 (see Stabilizing Interactions). Finally, the second helix of the C-terminal repeat was extended, since it is shorter by

Table 3. Mutated residues in the C-terminal cap of NI$_1$C

Mutant	Hydrophobic					Hydrophilic			α-Helix elongation		
	A83P	I86L	S87A	L95I	I98V	D89R	N90E	N92H	101K	102A	103A
Mut 1	X	X	X								
Mut 2				X	X						
Mut 3	X	X	X	X	X						
Mut 4									X	X	X
Mut 5	X	X	X	X	X				X	X	X
Mut 6	X	X	X	X	X	X	X	X	X	X	X

The consensus sequence[7] was used to suggest the type of side chain for each substitution. Additionally, the three-residue segment 101K–102A–103A was used to elongate the C-terminal helix. Two 300-K simulations of 40 ns and 50 ns with different initial assignments of velocities were performed for each mutant.

Results II

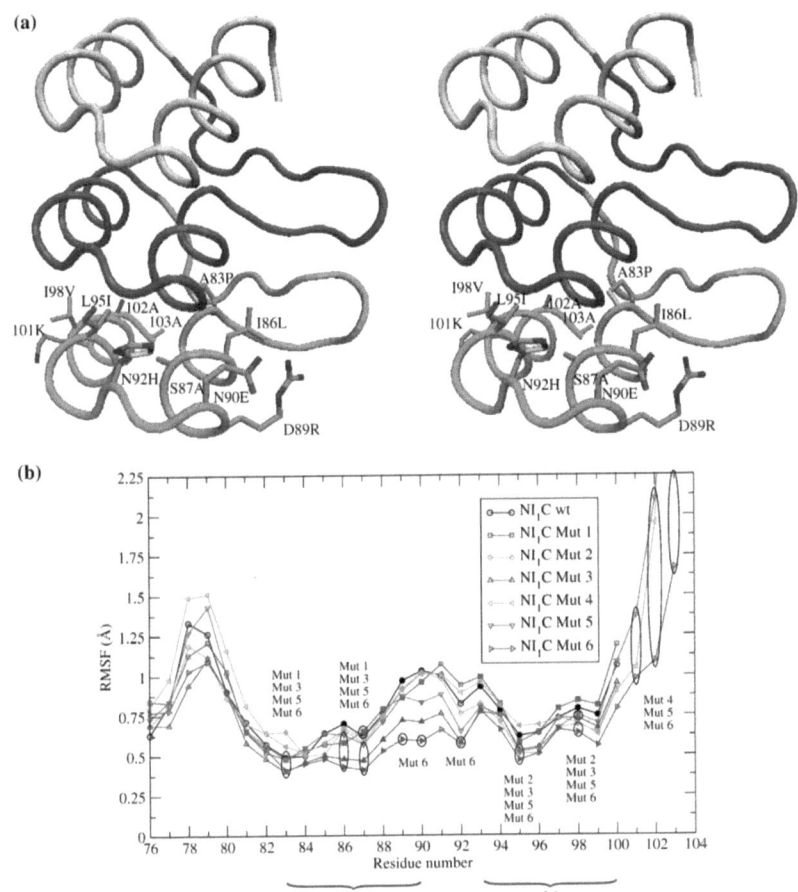

Fig. 8. (a) Stereo view of the NI₁C mutant with the eight point mutations and the three-residue C-cap extension 101K–102A–103A. (b) C_α RMSF of the C-terminal cap of wild-type NI₁C and the six mutant proteins, containing multiple substitutions (Table 3), at 300 K. The sites of mutation are enclosed by ellipses, and the names of the NI₁C mutant proteins that contain a substitution at the position indicated are given above or below the ellipse. The residues in helices are emphasized by filled symbols. The RMSF of the residue at position 103 is not displayed for Mut 4 and Mut 5. Their values are 4.54 Å and 3.44 Å, respectively. The last 40 ns of the 50-ns runs at 300 K were used to calculate the fluctuations. Quantitatively similar results are obtained using the 40-ns runs. A particular mutation is considered to increase the stability of the folded structure of NI₁C if the C_α RMSF at the site of the mutation is lower than that of the wild-type.

three residues than the consensus second helix of each repeat. The residues Lys-Ala-Ala were chosen for this extension because they increase the helical propensity.

Two main conclusions can be drawn from the 12 MD runs that represent an aggregated simulation time of 0.54 μs. First, the 11-point mutant of NI₁C (Mut 6) shows the lowest fluctuations (Fig. 8b) and the smallest number of different clusters of conformations (determined using the leader-clustering algorithm;[26,27] data not shown). This indicates that Mut 6 explores a more confined conformational space with respect to the other mutants and the wild-type. Second, individual hydrophobic side-chain replacements do not contribute significantly to the structural stability of the folded structure (Table 4). On the other hand, the mutations Asp89Arg and Asn90Glu together seem to be favorable, as they allow the Arg89–Glu90 salt bridge to form. In fact, the Arg89–Glu90 salt bridge is present in 84% of all frames sampled during the intervals 10–40 ns and 10–50 ns of the two 300-K simulations with Mut 6.

To test these suggestions, six mutants of NI_1C (see Supplementary Data), which contain the six multiple point mutations, were constructed (Table 3) in the C-terminal capping repeat. Expression in E. coli led to completely soluble proteins for all six mutants, and MALS analysis showed that all of the purified proteins are monomeric (data not shown). The stabilities of the six mutants were compared to the wild-type protein NI_1C using thermal denaturation experiments (Fig. 9a). While Mut 2 shows a $T_m = 57$ °C that is slightly below the transition midpoint of NI_1C wild-type (i.e., $T_m = 60$ °C), the other mutants show increased T_m values in the following order: Mut 4 with $T_m = 64$ °C, Mut 1 and Mut 3 with $T_m = 68$ °C, and Mut 5 and Mut 6 with 77 °C.

These results validate the hypotheses that replacing hydrophobic residues in the interrepeat interface with those present in the consensus sequence and elongation of the C-terminal helix can further improve the stability of DARPins. Indeed, Mut 1 and Mut 3, which present mutations of hydrophobic residues but no helix elongation, have an increased melting point of 8 degrees. On the other hand, Mut 4, which differs from the wild-type only by the elongation of the helix, shows a 4 degrees increase in stability. When mutations of hydrophobic residues and elongation of the C-terminal helix are combined, as in Mut 5 and Mut 6, an even larger increase in the melting point (17 degrees in total) is observed. However, mutations of hydrophobic residues in the C-terminal helix (i.e., L95I and I98V) cause a slight destabilization (the T_m of Mut 2 is 3 degrees less than for the wild-type), while these mutations have essentially no effect if they are combined with mutations in the first helix (Mut 3 and Mut 1 behave almost identically). Furthermore, the hypothesis of increased stability by additional electrostatic interactions (i.e., a salt bridge between the side chains at positions 89 and 90, and an interrepeat hydrogen bond involving a histidine at position 92) could not be confirmed. In fact, Mut 5 shows the same transition midpoint as Mut 6, although Mut 6 differs from Mut 5 by the additional mutations at positions 89, 90 and 92.

In summary, the stabilizing effect of the 11 mutations present in Mut 6 seems to be largely caused by six of them: the three present in Mut 1, which all bring the C-cap closer to the consensus (A83P, introducing a conserved Pro; I86L and S87A, increasing the helical propensity), as well as those present in Mut 4 (the extension of the second helix by Lys-Ala-Ala).

Mut 5 and Mut 6 were chosen for GdnHCl equilibrium unfolding measured by CD (Fig. 9b). The transitions of both mutants have the same midpoint, and they are cooperative, consistent with a two-state model. The calculated ΔG value (7.8 kcal/mol) is more than double that obtained for NI_1C wild-type (3.7 kcal/mol) and is 85% of the value for the four-repeat molecule NI_2C containing the wild-type C-cap.[19]

Because of the cooperative nature of NI_1C folding, this small protein serves well to quantify the effects of the stabilizing cap mutations. However, we also wished to test their contributions in the context of the larger NI_3C, to test their effect on the pretransition and the main transition. Therefore, in addition to introducing the C-cap mutations in NI_1C, they were also introduced in NI_3C. NI_3C Mut 5 and NI_3C Mut 6 are fully soluble, as are all the other mutants. However, NI_3C Mut 5 and NI_3C Mut 6 are a mixture of monomer and dimer (with about 15% dimer) at

Table 4. Side-chain contributions to the energy difference between Mut 6 and wild-type

	Side chain/system[a]		Side chain/protein[b]	
Mutation	Total	van der Waals	Total	van der Waals
Hydrophobic				
A83P	−3±2 (−5±1)	−4±1 (−4±1)	−6±2 (−4±1)	−8±1 (−4±1)
I86L	−1±1 (−13±1)	−1±1 (−12±1)	−3±1 (−10±1)	−2±1 (−10±1)
S87A	11±1 (−16±3)	1±1 (−6±1)	11±1 (−16±2)	0±1 (−5±1)
L95I	−1±1 (−13±1)	−1±1 (−12±1)	−1±1 (−12±1)	−1±1 (−11±1)
I98V	2±1 (−11±2)	−1±1 (−11±1)	−0±1 (−7±2)	1±1 (−8±1)
Polar				
D89R	58±40 (−156±46)	−11±2 (2±3)	−34±20 (−17±23)	−2±2 (−3±1)
N90E	−147±37 (−30±6)	6±3 (−7±2)	−52±24 (−9±4)	1±2 (−6±1)
N92H	−6±7 (−28±6)	−3±2 (−8±2)	−8±3 (−10±3)	−2±1 (−7±1)

All values are expressed in kilocalories per mole. A negative value indicates that the mutation is favorable. Values outside the parentheses are the energy difference between Mut 6 and wild-type, while values inside the parentheses are given inside the parentheses.
Only values of Mut 6 are shown (averages over the intervals 10–40 ns and 10–50 ns of the two 300-K simulations), as this mutant contains all substitutions. The results obtained with the other mutants for the same substitutions are not shown because their values fall within the standard deviation of those reported here for Mut 6.
[a] Total and van der Waals energies are calculated for a mutated side chain and the rest of the system (i.e., protein, ions and water), excluding the backbone of the mutated side chain.
[b] Total and van der Waals energies are calculated for a mutated side chain and the rest of the protein, excluding the backbone of the mutated side chain.

Results II

Discussion

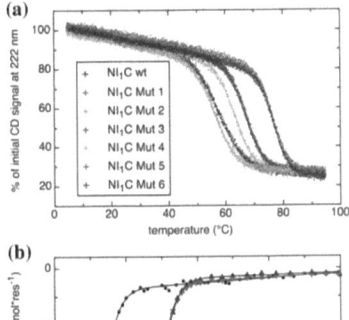

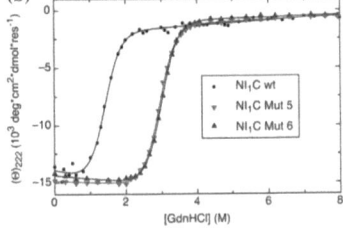

Fig. 9. Thermal and denaturant-induced unfolding of NI$_1$C mutants and wild-type followed by CD spectroscopy. (a) Thermal melting of NI$_1$C wild-type, NI$_1$C Mut 1, NI$_1$C Mut 2, NI$_1$C Mut 3, NI$_1$C Mut 4, NI$_1$C Mut 5 and NI$_1$C Mut 6 from 5 °C to 95 °C, using a heating gradient of 0.5 °C/min. Melting was only partially reversible. (b) GdnHCl-induced equilibrium unfolding of wild-type NI$_1$C, NI$_1$C Mut 5 and NI$_1$C Mut 6. The lines represent the least-squares fit to the two-state model with a midpoint of denaturation $D_m = 1.4$ ($\Delta G = 3.7 \pm 0.3$ kcal/mol) for wild-type, $D_m = 2.94$ M ($\Delta G = 7.8 \pm 0.2$ kcal/mol) for Mut 5 and $D_m = 3.0$ M ($\Delta G = 7.8 \pm 0.1$ kcal/mol) for Mut 6. The protein concentration was 10 μM in each case.

75 μM on MALS analysis. Under the assumption that the (presumably) even lower amount of dimer does not significantly influence the outcome of the experiment at the lower protein concentration of 10 μM, GdnHCl-induced equilibrium unfolding was measured by CD (Fig. 5a). In contrast to the unfolding curve of NI$_3$C wild-type, NI$_3$C Mut 5 and Mut 6 exhibit only a single transition at 6.7 M and at about 6.8 M GdnHCl, respectively. The increase in the D_m value is similar to the increase between NI$_1$C wild-type and NI$_1$C Mut 6. In thermal denaturation, it does not denature below 100 °C, but even wild-type NI$_3$C does not (Fig. 7).

Importantly, the pretransition in GdnHCl induced unfolding has disappeared in both NI$_3$C Mut 5 and Mut 6, fully consistent with our interpretation that this transition was due to the selective unfolding of the wild-type C-cap (Fig. 5). The more stable C-cap therefore "couples" to the rest of the protein, such that the main transition is moved to higher values. In summary, the weak link of the original C-cap (derived from the GA-binding protein) has been strengthened by our design.

Multiple MD simulations with explicit solvent were carried out to examine the stability and unfolding behavior of DARPins. These simulations totaled >2 μs and included the 300-K runs and high-temperature unfolding simulations. They involved designed repeat proteins of different lengths, comprising either "full"-consensus repeats or, in the case of the library member E3_5, consensus repeats differing in positions that had previously been randomized, as they are part of the potential binding interface.[7,10] In addition, variants without the capping repeats were examined, as were point mutants in the C-terminal capping repeat. To validate the simulations, experiments comparing the solubility, expression properties and equilibrium denaturation behavior of variants with and without capping repeats, as well as mutated capping repeats, were conducted.

Three main conclusions emerge from these studies. First, native-state stability appears to increase with the number of repeats, as can be seen by the time points of denaturation of the central repeats in the 400-K simulation (Fig. 3). Only the C-capping repeat denatures earlier (see below) and needs to be considered separately. This increase of stability with the number of repeats is consistent with the trend that has been observed in equilibrium denaturation experiments using either GdnHCl denaturation[12] or heat denaturation[7] with several members of the DARPin libraries. Similar results have also been reported from shortened variants of the ankyrin domains of the Drosophila Notch receptor or by insertion of consensus repeats into Notch.[16,24] Furthermore, the comparison of the previously described consensus sequences with two to four repeats also showed the same trends.[5,6]

As in all these previous experiments, either the sequences of the repeats were not identical or the proteins were not fully soluble; full-consensus proteins with caps identical with the ones simulated here have been constructed; and both equilibrium and kinetic stability were measured in GdnHCl and by temperature-induced unfolding, which are reported in detail in the accompanying manuscript.[19] The increase of stability with the increasing number of repeats was found with these identical (full-consensus) repeats as well.

Similar results were also found in studies of other repeat proteins. As an example, in a series of tetratrico peptide repeat proteins,[23] again a stability increase with the number of repeats was observed.

In the equilibrium unfolding experiments of our DARPins, we found either a single cooperative transition involving all repeats (as in the case of E3_5; Fig. 5 and a previous study[12]) or two transitions, with the first transition being interpreted as involving the C-terminal cap (see below) and with the main transition involving all other repeats (Fig. 5). In contrast, in force-induced unfolding experiments measured by atomic force microscopy, a sequential unfolding was observed.[28] This difference in unfolding behavior is not surprising, as force-induced unfolding imposes a particular direction on the unfolding trajectory. Furthermore, the force-induced unfolding is a kinetic experiment, while the solution experiments described here have been equilibrium experiments. A stepwise unfolding is observed in the high-temperature unfolding simulations described here. This is not at variance with experimental results. In the kinetic unfolding of a NX$_1$C library member protein, no intermediate was detected at 5 °C, but differential scanning calorimetry experiments revealed a deviation from a two-state model at higher temperatures.[29] Also, the interpretation of the equilibrium and kinetic unfolding data of the full consensus proteins[19] are consistent with folding models different from 2-state.

The sequence of unfolding events observed in the high-temperature simulations here is not at variance with Go-type simulations of the DARPin E3_5.[15] That study suggests that folding of E3_5 starts with the formation of the N-cap and propagates sequentially through neighboring repeats to the C-cap. However, in the unfolding simulation presented here, R3 unfolds after R2. This discrepancy could be due to the limited statistics (only one run with E3_5), the fact that unfolding might follow a slightly different pathway than folding or the presence of multiple folding pathways that are not detected by the simplified model used in that study.[15] Interestingly, that same study suggests that folding of the Notch receptor starts at the second or the sixth AR.[15] This disagrees with recent mutagenesis experiments showing that folding of the Notch receptor begins with the formation of repeats 3–5.[30] There, it is pointed out by the authors that the discrepancy between simulations and experiments of the Notch receptor could be due to the coarse-grained model used in the former.[30]

The second main conclusion derived from the present studies is that the full-consensus proteins show higher stability than even the most favorable library members. For example, E3_5, a member of the NX$_3$C library, can be compared with the full-consensus protein NI$_3$C. It should be noted that the stability of E3_5 is already very high, with a ΔG value of 11.2 ± 0.8 kcal/mol (determined by GdnHCl-induced equilibrium denaturation) and a melting temperature of >85 °C (Figs. 5 and 7). Nevertheless, further stability improvement can be observed in the full-consensus structure. Interestingly, the NI$_3$C molecule no longer shows a fully cooperative transition in GdnHCl-induced unfolding but a first transition where about 20% of the helical CD signal is lost. On the basis of the sequence of events observed in the MD simulations of unfolding, the first transition was interpreted as the loss of the C-cap structure (see below). Measurements with a protein lacking the C-cap, NI$_3$, supported this interpretation. The unfolding transition of NI$_3$ is cooperative, with a ΔG value of 23.6 ± 2.4 kcal/mol, and occurs at the same GdnHCl concentration as the second transition (I⇌U) for NI$_3$C (Fig. 5a). Although NI$_3$ aggregates in GdnHCl at higher protein concentrations, it was mainly monomeric at 7 µM, as shown by analytical ultracentrifugation (Fig. 5c). A ΔG value of 19.7 ± 4.6 kcal/mol was calculated from the experimental equilibrium unfolding data for NI$_3$C using a sequential three-state model.[31] The presence of a third species in equilibrium unfolding has been observed as well for the natural ankyrin p19;[32] however, in this protein, several repeats strongly deviate from the consensus sequence and might constitute a "weak link." Furthermore, this protein is substantially less stable ($D_{m,urea} = 2.9$ M). GdnHCl-induced equilibrium unfolding experiments with NI$_4$ showed no transition, indicating an even higher stability for a protein with four full-consensus repeats. However, the unfolding study with this protein is difficult, as it is very prone to aggregation in GdnHCl, as also observed with NI$_3$ in GdnHCl.

In designing this "full-consensus" sequence, those residues that mediate binding to target proteins in DARPins[7,10] were replaced by the most frequent residues,[19] and thereby a number of charged residues were newly introduced. These are involved in additional salt bridges (e.g., between β-hairpins), and they are likely to contribute to the unusual stability.

The third main conclusion is that the C-cap used here is the limiting part for the stability of the whole-consensus repeat protein. Note, however, that this becomes only experimentally noticeable in the most extremely stable molecules. In the majority of library members, which still have ΔG values of 10–20 kcal/mol measured in equilibrium denaturation experiments and melting temperatures of between 70 and 90 °C[7,12] and are thus already at the upper edge of natural proteins, a single cooperative transition characterizing obviously very stable molecules is found. This indicates that the engineering of the C-cap will be of importance only for applications under the most extreme conditions and might push the already highly stable DARPins even further.

As mentioned above, the C-cap denatured first in almost all MD runs of the designed consensus ARs. A possible reason is that the C-cap originates from the natural GA-binding protein and has not been under particular evolutionary pressure. Thus, its amino acid sequence significantly differs from the full-consensus design and is characterized by a shorter helix. These differences possibly lead to a low structural stability of the shorter helix, to a poor packing of the hydrophobic core at the interface to the preceding repeat and to the lack of one interrepeat hydrogen bond and one intrarepeat salt bridge. The last two electrostatic interactions are indeed present among consensus repeats. These observations were taken into account to suggest the design of an even more stable C-terminal capping repeat (see below).

The C-capping repeat plays a very important role: In its absence, the expression of DARPins in *E. coli* leads to a significant amount of insoluble aggregated protein. The C-capping repeat prevents formation of insoluble aggregates and the N-capping repeat prevents soluble aggregates, while the con-

Results II

structs I_3 and I_4 (missing both capping repeats) are expressed almost entirely in inclusion bodies (Fig. 6). These observations explain the evolution of the capping repeats to secure the cellular folding and function of AR proteins and also demonstrate the importance of the capping repeats for the practical utilization of DARPins in biotechnology.

These findings are also consistent with the report on another design study of full-consensus AR proteins[5] with a slightly different sequence and without any caps, which were only soluble at acidic pH. The introduction of positive charges in the C-cap then allowed the protein to be soluble at neutral pH, but it still had to be produced from inclusion bodies made in *E. coli* with subsequent refolding.[6] However, the gain in solubility was accompanied by a significant loss in stability at pH 4. Furthermore, the stability could not be measured at pH 7 because the protein was not soluble.

The C-cap has thus been identified as being absolutely necessary to provide a highly charged surface to the protein to allow it to fold to the native state in a bacterial expression system, but at the same time to become a liability if one wants to drive the stability of these proteins to even more extreme values. The combined simulation and experimental results led to the question on whether it might be possible to design an equally soluble C-cap that nevertheless was of a similar stability as the internal consensus repeats. Eight different point mutations were considered, as was an extension of three amino acids to the last helix. The variant containing all mutations, as well as the C-terminal extension (NI_1C Mut 6), showed significantly smaller fluctuations than the wild-type in room-temperature MD simulations (Fig. 8b).

Testing the mutations in equilibrium unfolding experiments largely confirmed the suggested design. All the six mutants are equally soluble as the wild-type protein. Both NI_1C Mut 5 and NI_1C Mut 6 show a remarkable increase in stability, as indicated by a melting point that lies 17 degrees higher and by a >2-fold increase in ΔG value when compared to the wild-type NI_1C. These results confirmed the importance of a better packing of the hydrophobic core. However, the introduction of additional electrostatic interactions does not further increase the stability, as indicated by the similar curves of Mut 5 and Mut 6 (Figs. 5a and 9). When the eight point mutations and the three-residue helical extension are introduced into NI_3C (NI_3C Mut 5 and Mut 6), we also observe a large increase in stability compared to NI_3C wild-type, but more importantly, the pretransition at 3.7 M GdnHCl is absent (Fig. 5a). This experiment is further proof that the equilibrium intermediate in NI_3C wild-type corresponds to a state wherein the less stable wild-type C-terminal capping repeat is selectively denatured.

The current study has a number of direct implications for the understanding of repeat proteins and the design of further improved libraries. It helps to rationalize the minimal number of ARs found in natural proteins, as the critical interactions between repeats are important for stabilizing the repeat domain. It also helps to understand the vital importance of the capping repeats and shows that if extreme stabilities are needed, the C-cap of GA-binding protein can become limiting. However, with the improved design of the C-cap, DARPins of even more extreme stability can be designed.

Materials and Methods

Sequences and initial conformations

Table 1 lists the systems that were simulated in the present study. Simulations of E3_5 were started from its X-ray structure[12] (PDB code 1MJ0). The initial conformations of NI_1C, NI_2C and NI_3C were modeled from the structure of E3_5. The mutated side chains were constructed with a library of rotamers using the program Insight II (Accelrys, Inc.). The experimental structure of NI_3C has meanwhile been determined and is described in the accompanying paper.[33]

The first and last residues of each repeat are defined here differently from Refs. 7, 12 and 34 for topological reasons. In the present work, each internal repeat includes six amino acids preceding the β-hairpin at the tip of the loops, while in Refs. 7, 12 and 34, that β-hairpin was used as the start (Supplementary Fig. 1). The present position Ala1 of each repeat would correspond to Ala28 of the previous repeat in Refs. 7, 12 and 34. In this way, intra-loop contacts (such as hydrogen bonds and salt bridges) being counted as interrepeat contacts is avoided. This definition is also used to calculate repeatwise RMSDs from the initial conformation.

For the "full-consensus" AR proteins NI_xC, the variable positions in Ref. 7 were fixed.[19] In the notation of this manuscript, the primary structure of the full-consensus internal repeats of NI_xC is $A_1DVNAKD$ KDG_{10}-YTPLHLAARE$_{20}$ GHLEIVEVLL$_{30}$KAG$_{33}$, where the newly defined residues that differ in E3_5 and other members of the library are in boldface (cf. Supplementary Fig. 1). For the discussion of E3_5 and NI_3C, the residue number refers to the numbering scheme according to PDB file 1MJ0. The constructs missing the N-cap (I_1, I_2, I_3, I_4, I_3C, and I_4C) start in front of the first helix of the internal-consensus repeat (sequence TPLHL, position 12 of the numbering scheme shown above, corresponding to position 49 in PDB file 1MJ0; see also Supplementary Fig. 1). The constructs missing the C-cap (I_1, I_2, I_3, I_4, NI_3 and NI_4) end with the second helix of the internal repeat (sequence LLKAG, position 33 of the numbering scheme shown above, corresponding to position 136 in the PDB file of E3_5; PDB code 1MJ0). This molecule, E3_5, has the same length as an NI_3C molecule (see also Supplementary Fig. 1).

For the simulations, the C-terminal cap of the NI_1C mutants was modeled by homology by superimposing the central repeat of NI_1C to the C-terminal cap to generate the coordinates of the mutated side chains and the three-amino-acid extension as well.

Simulations

The MD simulations were performed with the program NAMD2[35] using the CHARMM all-hydrogen force field (PARAM22)[36] and the TIP3P model of water. To effectively compare simulations with experimental results (e.g., a pH

Results II

of 7.4 in the CD experiments; see below), side chains of aspartates and glutamates were negatively charged, those of lysines and arginines were positively charged and histidines were considered neutral. Initial conformations were minimized in vacuo by performing 100 steps of steepest descent and subsequently 500 steps of conjugate gradient minimization with CHARMM.[37] The proteins were then inserted into a cubic water box of different side lengths, depending on the number of amino acids. In the case of NI_3C and E3_5, a larger box was used for the 400-K than for the 300-K simulations. The minimal distance between the protein and the boundary of the box was 12 Å. The different box sizes and durations of the simulations are summarized in Table 1. Furthermore, for each of the six mutants of NI_1C, two simulations at 300 K (50 ns and 40 ns) were performed using a box with the same dimensions as the one used for NI_1C. Chloride and sodium ions were added to neutralize the system and approximate a salt concentration of 150 mM. The water molecules overlapping with the protein or the ions were removed if the distance between the water oxygen and any atom of the protein or any ion was smaller than 3.1 Å. The number of water molecules ranged from 3906 to 31,443, and the total number of atoms ranged between 12,087 and 96,878. To avoid finite-size effects, periodic boundary conditions were applied. Different initial random velocities were assigned whenever more than one simulation was performed with the same protein. Electrostatic interactions were calculated within a cutoff of 12 Å, while long-range electrostatic effects were taken into account by the Particle Mesh Ewald summation method.[38] Van der Waals interactions were treated with the use of a switch function starting at 10 Å and turning off at 12 Å. The temperature was kept constant by using the Berendsen thermostat[39] with a relaxation time of 0.2 ps, while the pressure was held constant at 1 atm by applying a pressure piston.

Before production runs, harmonic constraints were applied to the positions of all the atoms of the protein to equilibrate the system at 300 K or 400 K during a time length of 0.2 ns. After this equilibration phase, the harmonic constraints were released. For the runs at 300 K, the first 10 ns of unconstrained simulation time were also considered part of the equilibration and were thus not used for the analysis. For the six mutants of NI_1C, the equilibration was elongated by 2 ns without restraints on the mutated amino acids and its two neighbors. The dynamics was integrated with a time step of 2 fs. The covalent bonds involving hydrogens were rigidly constrained by means of the SHAKE algorithm with a tolerance of 10^{-8}. Snapshots were saved every 2 ps for trajectory analysis.

Determination of native contacts

The conformations sampled at room temperature were used to determine native hydrogen bonds, salt bridges and C_α contacts. To define a hydrogen bond, a H⋯O distance cutoff of 2.7 Å and a D–H⋯O angle cutoff of 120° were used, where a donor D could either be an oxygen or a nitrogen. An interaction was defined as a salt bridge if the N_ζ of Lys or the C_ζ of Arg was closer than 4 Å or 5 Å, respectively, from either the C_γ of Asp or the C_δ of Glu. All histidines were assumed to be neutral. A C_α contact involves two C_α atoms with a distance smaller than 6.5 Å and not adjacent in sequence (i.e., residue pairs i,j, $j > i+2$). Only those hydrogen bonds, salt bridges and C_α contacts present in at least half of the simulation frames at 300 K were selected as native contacts (Table 2). They were used to compare the conformational flexibility of different proteins at room temperature and to monitor the changes in secondary and tertiary structures during unfolding.

Design and synthesis of DNA-encoding AR proteins, protein expression and purification

The process of the sequence design of the full-consensus ARs is described in the accompanying manuscript.[19] The cloning, expression and purification of DARPins have been performed as described elsewhere.[7] The construction of the C-cap mutants is described in the Supplementary Data.

CD spectroscopy

All CD experiments were performed in 50 mM sodium phosphate buffer (pH 7.4) and 150 mM NaCl, using 5–10 μM protein purified by immobilized metal ion affinity chromatography as described.[7,10] To measure the denaturant-induced equilibrium unfolding curves, the samples were equilibrated at 20 °C overnight at the corresponding GdnHCl concentrations. The CD signal at 222 nm was recorded on a Jasco J-715 instrument (Jasco, Japan) equipped with a computer-controlled water bath, using a cylindrical quartz cell of 1 mm path length. CD data were collected at 222 nm and 20 °C every 5 s with a bandwidth of 2 nm and a response time of 4 s, averaged over 2 min. A baseline correction was made with the buffer. The CD signal was converted to mean residue ellipticity (Θ_{MRE}) using the concentration of the sample determined spectrophotometrically at 280 nm. Thermal unfolding was recorded by continuous heating from 5 °C to 95 °C with a temperature gradient of 0.5 °C/min. CD data were collected at 222 nm every 5 s with a bandwidth of 2 nm and a response time of 4 s. Reversibility was determined from the recovery of ellipticity after cooling.

Fluorescence spectroscopy

Tyrosine fluorescence was excited at 274 nm, and emission spectra were recorded from 290 to 350 nm using a PTI Alpha Scan spectrofluorimeter (Photon Technologies, Inc.). Slid widths of 5 nm were used for both excitation and emission. Samples were prepared as for the CD measurements. After buffer correction, the intensity of the emission maximum at 304 nm or 303 nm, respectively, was plotted against the denaturant concentration.

Analytical ultracentrifugation

Sedimentation equilibrium experiments were performed with a Beckman XL-A centrifuge with a NA-50 Ti rotor at 20 °C using optical absorbance detection. NI_3 protein at two concentrations (0.1 mg/ml and 0.3 mg/ml) was measured in 2 M GdnHCl, 50 mM phosphate buffer and 150 mM NaCl (pH 7.4). Protein and buffer samples were placed in cells fitted with double-sector centerpieces and quartz windows. Sedimentation equilibrium was approached at a rotor speed of 35,000 rpm. The cells were scanned at 230 nm, and 40 scans were collected. The scans were analyzed using the software SEDPHAT‡.

---

‡ http://www.analyticalultracentrifugation.com/sedphat/ (by P. Schuck, National Institutes of Health, Bethesda, MD).

Results II

Solvent density was calculated from the weight of the salts, and the partial specific volume of the protein was determined from the amino acid sequence of the protein using the software UltraScan§, not taking into account the influence of dissolved GdnHCl on the partial specific volume.

Acknowledgements

We thank Dr. H.K. Binz for preparing the homology model of NI$_3$C, and Dr. C. Bodenreider (Universität Basel) for help with data fitting. We are grateful to Dr. C. Briand for valuable help with analytical ultracentrifugation, and Dr. P. Forrer for helpful discussions about library design and biophysical properties of AR proteins. We thank A. Widmer (Novartis Pharma, Basel, Switzerland) for providing the molecular modeling program WITNOTP, which was used for visual analysis of the trajectories. The simulations were performed on the Matterhorn Beowulf cluster at the Computing Center of the University of Zürich. We thank C. Bolliger and Dr. A. Godknecht for setting up the cluster, and the Canton of Zürich for generous hardware support. This work was supported by a Swiss National Science Foundation grant to A.C. and by a National Center of Competence in Research in Structural Biology grant to A.P.

Supplementary Data

Supplementary data associated with this article can be found, in the online version, at doi:10.1016/j.jmb.2007.09.042

References

1. Andrade, M. A., Perez-Iratxeta, C. & Ponting, C. P. (2001). Protein repeats: structures, functions, and evolution. *J. Struct. Biol.* **134**, 117–131.
2. Kobe, B. & Kajava, A. V. (2000). When protein folding is simplified to protein coiling: the continuum of solenoid protein structures. *Trends Biochem. Sci.* **25**, 509–515.
3. Groves, M. R. & Barford, D. (1999). Topological characteristics of helical repeat proteins. *Curr. Opin. Struct. Biol.* **9**, 383–389.
4. Bork, P. (1993). Hundreds of ankyrin-like repeats in functionally diverse proteins: mobile modules that cross phyla horizontally? *Proteins: Struct. Funct. Bioinf.* **17**, 363–374.
5. Mosavi, L. K., Minor, D. L. & Peng, Z. Y. (2002). Consensus-derived structural determinants of the ankyrin repeat motif. *Proc. Natl Acad. Sci. USA*, **99**, 16029–16034.

§ http://www.ultrascan.uthscsa.edu/ (by B. Demeler, University of Texas Health Science Center, San Antonio, TX).

6. Mosavi, L. K. & Peng, Z. Y. (2003). Structure-based substitutions for increased solubility of a designed protein. *Protein Eng.* **10**, 739–745.
7. Binz, H. K., Stumpp, M. T., Forrer, P., Amstutz, P. & Plückthun, A. (2003). Designing repeat proteins: well-expressed, soluble and stable proteins from combinatorial libraries of consensus ankyrin repeat proteins. *J. Mol. Biol.* **332**, 489–503.
8. Forrer, P., Stumpp, M. T., Binz, H. K. & Plückthun, A. (2003). A novel strategy to design binding molecules harnessing the modular nature of repeat proteins. *FEBS Lett.* **539**, 2–6.
9. Batchelor, A. H., Piper, D. E., de la Brousse, F. C. & McKnight, S. L. (1998). The structure of GABP alpha/beta: an ETS domain ankyrin repeat heterodimer bound to DNA. *Science*, **279**, 1037–1041.
10. Binz, H. K., Amstutz, P., Kohl, A., Stumpp, M. T., Briand, C., Forrer, P. *et al.* (2004). High-affinity binders selected from designed ankyrin repeat protein libraries. *Nat. Biotechnol.* **22**, 575–582.
11. Amstutz, P., Binz, H. K., Parizek, P., Stumpp, M. T., Kohl, A., Grütter, M. G. *et al.* (2005). Intracellular kinase inhibitors selected from combinatorial libraries of designed ankyrin repeat proteins. *J. Biol. Chem.* **280**, 24715–24722.
12. Kohl, A., Binz, H. K., Forrer, P., Stumpp, M. T., Grütter, M. G. & Plückthun, A. (2003). Designed to be stable: crystal structure of a consensus ankyrin repeat protein. *Proc. Natl Acad. Sci. USA*, **100**, 1700–1705.
13. Tang, K. S., Fersht, A. R. & Itzhaki, L. S. (2003). Sequential unfolding of ankyrin repeats in tumor suppressor p16. *Structure*, **11**, 67–73.
14. Interlandi, G., Settanni, G. & Caflisch, A. (2006). Unfolding transition state and intermediates of the tumor suppressor p16^{ink4a} investigated by molecular dynamics simulations. *Proteins: Struct. Funct. Bioinf.* **64**, 178–192.
15. Ferreiro, D. U., Cho, S. S., Komives, E. A. & Wolynes, P. G. (2005). The energy landscape of modular repeat proteins: topology determines folding mechanism in the ankyrin family. *J. Mol. Biol.* **354**, 679–692.
16. Mello, C. C. & Barrick, D. (2004). An experimentally determined protein folding energy landscape. *Proc. Natl Acad. Sci. USA*, **101**, 14102–14107.
17. Tripp, K. W. & Barrick, D. (2004). The tolerance of a modular protein to duplication and deletion of internal repeats. *J. Mol. Biol.* **344**, 169–178.
18. Mello, C. C., Bradley, C. M., Tripp, K. W. & Barrick, D. (2005). Experimental characterization of the folding kinetics of the notch ankyrin domain. *J. Mol. Biol.* **352**, 266–281.
19. Wetzel, S. K., Settanni, G., Kenig, M., Binz, H. K. & Plückthun, A. (2007). Folding and unfolding mechanism of highly stable full consensus ankyrin repeat proteins. *J. Mol. Biol.* In press. doi:10.1016/j.jmb.11.046.
20. Karshikoff, A. & Ladenstein, R. (2001). Ion pairs and the thermotolerance of proteins from hyperthermophiles: a 'traffic rule' for hot roads. *Trends Biochem. Sci.* **26**, 550–556.
21. Berezovsky, I. N. & Shakhnovich, E. I. (2005). Physics and evolution of thermophilic adaptation. *Proc. Natl Acad. Sci. USA*, **102**, 12742–12747.
22. Zhang, B. & Peng, Z.-Y. (2000). A minimum folding unit in the ankyrin repeat protein p16(ink4). *J. Mol. Biol.* **299**, 1121–1132.
23. Main, E. R. G., Stott, K., Jackson, S. E. & Regan, L. (2005). Local and long-range stability in tandemly

arrayed tetratricopeptide repeats. *Proc. Natl Acad. Sci. USA*, **102**, 5721–5726.
24. Tripp, K. W. & Barrick, D. (2007). Enhancing the stability and folding rate of a repeat protein through the addition of consensus repeats. *J. Mol. Biol.* **26**, 1187–1200.
25. Binz, H. K., Kohl, A., Plückthun, A. & Grütter, M. G. (2006). Crystal structure of a consensus-designed ankyrin repeat protein: implications for stability. *Proteins: Struct. Funct. Bioinf.* **65**, 280–284.
26. Hartigan, J. A. (1975). *Clustering Algorithms*. Wiley, New York.
27. Settanni, G., Rao, F. & Caflisch, A. (2005). Φ-Value analysis by molecular dynamics simulations of reversible folding. *Proc. Natl Acad. Sci. USA*, **102**, 628–633.
28. Li, L. W., Wetzel, S., Plückthun, A. & Fernandez, J. M. (2006). Stepwise unfolding of ankyrin repeats in a single protein revealed by atomic force microscopy. *Biophys. J.* **90**, L30–L32.
29. Devi, V. S., Binz, H. K., Stumpp, M. T., Plückthun, A., Bosshard, H. R. & Jelesarov, I. (2004). Folding of a designed simple ankyrin repeat protein. *Protein Sci.* **13**, 2864–2870.
30. Bradley, C. M. & Barrick, D. (2006). The notch ankyrin domain folds via a discrete, centralized pathway. *Structure*, **14**, 1303–1312.
31. Barrick, D. & Baldwin, R. L. (1993). Three-state analysis of sperm whale apomyoglobin folding. *Biochemistry*, **32**, 3790–3796.
32. Zeeb, M., Rosner, H., Zeslawski, W., Canet, D., Holak, T. A. & Balbach, J. (2002). Protein folding and stability of human cdk inhibitor p19(INK4d). *J. Mol. Biol.* **315**, 447–457.
33. Merz, T., Wetzel, S. K., Firbank, S., Plückthun, A., Grütter, M. G. & Mittl, P. R. E. (2007). Stabilizing ionic interactions in a full consensus ankyrin repeat protein. *J. Mol. Biol.* In press. doi:10.1016/j.jmb.11.047.
34. Sedgwick, S. G. & Smerdon, S. J. (1999). The ankyrin repeat: a diversity of interactions on a common structural framework. *Trends Biochem. Sci.* **24**, 311–316.
35. Kalé, L., Skeel, R., Bhandarkar, M., Brunner, R., Gursoy, A., Krawetz, N. *et al.* (1999). NAMD2: greater scalability for parallel molecular dynamics. *J. Comp. Phys.* **151**, 283–312.
36. MacKerell, A. D. E. A., Jr (1998). All-atom empirical potential for molecular modeling and dynamics studies of proteins. *J. Phys. Chem. B*, **102**, 3586–3616.
37. Brooks, B. R., Bruccoleri, R. E., Olafson, B. D., States, D. J., Swaminathan, S. & Karplus, M. (1983). CHARMM: a program for macromolecular energy, minimization, and dynamics calculations. *J. Comput. Chem.* **4**, 187–217.
38. Darden, T., York, D. & Pedersen, L. (1993). Particle Mesh Ewald—an $N \cdot \log(N)$ method for Ewald sums in large systems. *J. Chem. Phys.* **98**, 10089–10092.
39. Berendsen, H. J. C., Postma, J. P. M., Van Gunsteren, W. F., Dinola, A. & Haak, J. R. (1984). Molecular-dynamics with coupling to an external bath. *J. Chem. Phys.* **81**, 3684–3690.
40. Humphrey, W., Dalke, A. & Schulten, K. (1996). VMD: visual molecular dynamics. *J. Mol. Graphics*, **14**, 33–38.

Results **III**

2.3 Stabilizing Ionic Interactions in a Sulfate Binding Ankyrin Repeat Protein (***III***)

Here, the crystal structure of the full consensus DAPRin NI_3C is determined and compared to two N3C library members. Further considerations are made to explain the high stability of the full consensus AR.

Merz, T., Wetzel, S. K., Firbank, S., Plückthun, A., Grütter, M. & Mittl, P. R. E. (2008). Stabilizing ionic interactions in a full consensus ankyrin repeat protein. *J. Mol. Biol.* **376**, 232-240
(for the coloured version of the article refer to the publisher weblink
http://dx.doi.org/10.1016/j.jmb.2007.11.047)

III

Stabilizing Ionic Interactions in a Full-consensus Ankyrin Repeat Protein

Tobias Merz, Svava K. Wetzel, Susan Firbank, Andreas Plückthun, Markus G. Grütter and Peer R. E. Mittl*

Department of Biochemistry, University of Zürich, Winterthurerstrasse 190, CH-8057 Zürich, Switzerland

Received 4 July 2007; received in revised form 28 September 2007; accepted 16 November 2007
Available online 22 November 2007

Full-consensus designed ankyrin repeat proteins (DARPins), in which randomized positions of the previously described DARPin library have been fixed, are characterized. They show exceptionally high thermodynamic stabilities, even when compared to members of consensus DARPin libraries and even more so when compared to naturally occurring ankyrin repeat proteins. We determined the crystal structure of a full-consensus DARPin, containing an N-capping repeat, three identical internal repeats and a C-capping repeat at 2.05 Å resolution, and compared its structure with that of the related DARPin library members E3_5 and E3_19. This structural comparison suggests that primarily salt bridges on the surface, which arrange in a network with almost crystal-like regularity, increase thermostability in the full-consensus NI_3C DARPin to make it resistant to boiling. In the crystal structure, three sulfate ions complement this network. Thermal denaturation experiments in guanidine hydrochloride directly indicate a contribution of sulfate binding to the stability, providing further evidence for the stabilizing effect of surface-exposed electrostatic interactions and regular charge networks. The charged residues at the place of randomized residues in the DARPin libraries were selected based on sequence statistics and suggested that the charge interaction network is a hidden design feature of this protein family. Ankyrins can therefore use design principles from proteins of thermophilic organisms and reach at least similar stabilities.

© 2007 Elsevier Ltd. All rights reserved.

Edited by F. Schmid

Keywords: repeat protein; protein stability; salt bridge; thermodynamic stability; X-ray crystallography

Introduction

Repeat proteins consist of repeating structural units that stack together to form elongated non-globular domains.[1,2] In contrast to globular proteins, they are not stabilized by interactions between residues that are very distant in sequence; instead, the stabilizing and structure-determining interactions are formed within a repeat and between neighboring repeats. Repeat proteins can be extended in size while still constituting a contiguous domain, making them unique targets for protein engineering. Repeat proteins constitute, next to immunoglobulins, the most abundant natural protein classes specialized in binding.

Because of their abundance and the multiple occurrences of repeats within one protein sequence, a statistical analysis of thousands of sequences can be carried out to design consensus repeats. This has been reported for ankyrin repeat (AR) proteins, tetratricopeptide repeat proteins and leucine-rich repeat proteins.[3–6] The AR is one of the most common protein sequence motifs. This 33-residue motif consists of a β-turn, followed by two antiparallel α-helices and a loop reaching the turn of the next repeat.[7]

*Corresponding author. E-mail address: mittl@bioc.uzh.ch.
Present address: S. Firbank, Institute of Cell and Molecular Biosciences, University of Newcastle, Framlington Place, Newcastle upon Tyne NE2 4HH, UK.
Abbreviations used: DARPin, designed ankyrin repeat protein; AR, ankyrin repeat; PDB, Protein Data Bank; SC, shape complementarity; HB, hydrogen bonds; Gdn·HCl, guanidinium hydrochloride.

0022-2836/$ - see front matter © 2007 Elsevier Ltd. All rights reserved.

Two independent approaches used to apply consensus design strategies to AR proteins have been reported so far.[3,4] Both employ a similar but not identical consensus version of the 33-residue AR.[7] In one approach, large libraries of AR proteins,[3] in which only 26 of the 33 amino acids were specified while 7 were allowed to vary in order to bind to different target molecules, were made.[8]

The potential target interaction residues of a single AR (2, 3, 5, 13, 14 and 33), henceforth named randomized residues, are located on the concave side of designed ankyrin repeat proteins (DARPins).[3] Highly specific binders for a number of different target proteins have been selected using the DARPin libraries[8], and structures of ankyrin–target protein complexes have been determined.[9–12] In this approach, N- and C-capping repeats flanking the randomized repeats that shield the hydrophobic core were employed. Recently, it was shown experimentally that indeed the capping repeats are required for soluble expression in *Escherichia coli*[13] and for shielding of the protein against aggregation.

In another approach, a full consensus was derived from sequence statistics.[4] In these studies, no capping repeats were employed, and the resulting proteins were only soluble at acidic pH.[4] The introduction of positive charges in the C-terminal consensus repeat then allowed the protein to be soluble at neutral pH, but it still had to be produced by *in vitro* refolding from inclusion bodies made in *E. coli*.[5] However, the gain in solubility was accompanied by a significant loss in stability at pH 4.

As described in detail elsewhere in this issue,[14] a full-consensus version of the DARPins, in which randomized residues of the library have now been defined such that every internal repeat has exactly the same sequence, was designed. We denote these proteins as NI$_x$C, where "N" and "C" refer to the N- and C-terminal capping repeats, respectively, "I" refers to the internal full-consensus repeat and the subscript x gives the number of identical internal consensus repeats. The design of NI$_x$C proteins and their thermodynamic and kinetic folding properties have been investigated.[14] The N- and C-terminal capping repeats were originally taken[3] from GABP β1 (guanine-adenine binding protein β1) (PDB code 1AWC).[15]

Here we report the crystal structure of the full-consensus DARPin NI$_3$C at 2.05 Å resolution and compare its features with the structures of two related consensus DARPins E3_5 and E3_19 [Protein Data Bank (PDB) entries 1MJ0 and 2BKG, respectively]. All three molecules were designed using the same framework residues. In contrast to NI$_3$C, the internal repeats of E3_5[16] and E3_19[17] do not have the same sequences because they contain different residues at the randomized positions. From our structural analysis, we propose that a highly regular array of salt bridges, the overall charge distribution and the binding of sulfate ions significantly contribute to the thermodynamic stability of NI$_3$C. This study is intended to extend our understanding of stabilizing charge effects in AR proteins.

Results

Overall structure

The final model of the full-consensus NI$_3$C DARPin structure at 2.05 Å resolution contains all amino acids and side chains of the three identical internal repeats and the flanking N- and C-terminal capping repeats. The capping repeats expose a hydrophilic surface and shield the hydrophobic core of the internal repeats from the solvent to prevent aggregation. As in all AR proteins, the repeats adopt an L-shaped arrangement, which is caused by two almost antiparallel α-helices and a β-turn forming the interrepeat connection (Fig. 1a). Residues positioned at the concave side of the DARPin library (randomized positions) normally mediate specific interactions with selected targets. Based on the statistics of naturally occurring AR protein sequences, they were designed to be highly charged in the full-consensus version.[14] As a consequence, three sulfate ions from the crystallization buffer were found to bind to this site (Fig. 1b).

E3_5/E3_19 consensus *versus* NI$_3$C full consensus

Sequence differences within repeats between the full-consensus molecule NI$_3$C and the unselected library members E3_5 and E3_19 occur at positions 2, 3, 5, 13, 14, 26 and 33 (for repeat numbering, see Fig. 2). Residues at these randomized positions depend on the selection process and typically differ from one repeat to the next. In contrast to E3_5 and E3_19, all internal repeats in NI$_3$C are identical (Fig. 2). As discussed in detail elsewhere in this issue,[14] amino acid types were selected by sequence statistics and structural considerations.

Briefly, at positions 2 and 3, lysine and aspartic acid residues were introduced, respectively. While lysine is the most prevalent amino acid at position 2, aspartic acid is the second most abundant amino acid at position 3 (after asparagine). A negatively charged amino acid was selected for position 3 to balance the positive charges that were introduced at positions 2 and 33. For position 5, which is quite variable, tyrosine was introduced as a spectroscopic probe. In position 13, the second most abundant arginine (after glutamine) was used to compensate for negative charges, and the most abundant glutamic acid was selected for position 14. Alanine was inserted at position 26 because it is most abundant at this position and possesses high helical propensity. For position 33, the most abundant lysine was selected. Figure 2 shows the correspondence between the numbering within a single repeat and the protein sequence.

Results III

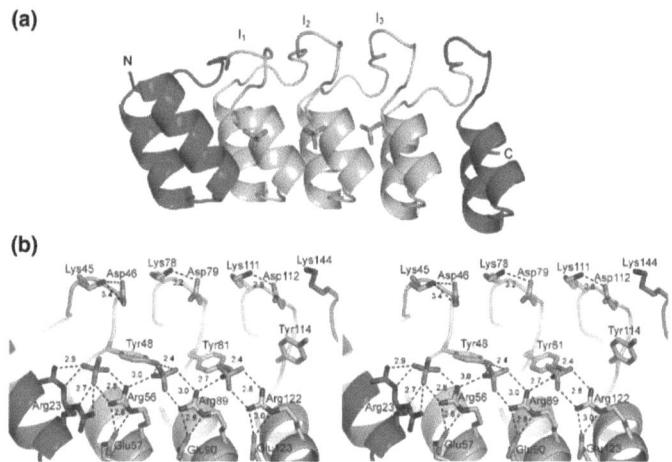

Fig. 1. (a) Ribbon diagram of the NI$_3$C structure. The terminal and internal repeats are in magenta and green, respectively. (b) Stereo view of the interactions involving the three sulfate ions and the randomized residues. Dashed lines in black denote HB between lysine and aspartate residues at positions 2 and 3, as well as HB between arginine and glutamate residues at positions 13 and 14. Dashed lines in blue denote all interactions of the sulfate ions.

Structural comparison with other AR proteins

The overall structures of NI$_3$C, E3_5 and E3_19 are very similar. The RMSD values of the pairwise comparisons of Cα atoms between NI$_3$C/E3_5, NI$_3$C/E3_19 and E3_5/E3_19 are 0.62 Å, 0.50 Å and 0.60 Å, respectively. Superpositions of the three molecules NI$_3$C, E3_5 and E3_19 revealed the largest structural differences in the C-terminal capping repeats, which move, in a first approximation, as rigid bodies. This is illustrated by a significant reduction in the overall RMSD values when omitting

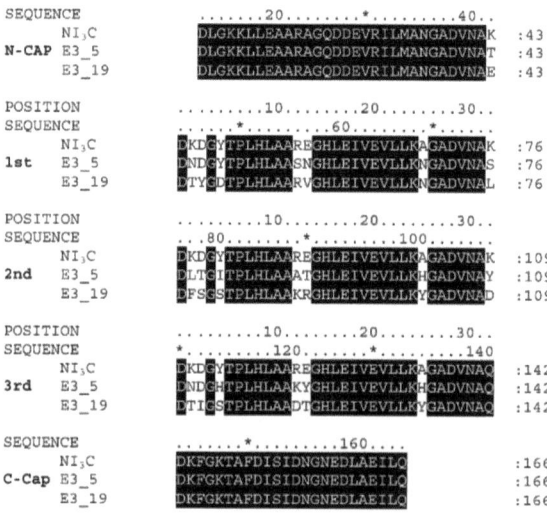

Fig. 2. Sequence alignment and numbering of NI$_3$C, E3_5 and E3_19. The N-terminal capping repeat includes a His$_6$ tag (MRGSHHHHHHGS; sequence not shown). Identical residues are shown against a black background, whereas randomized residues at positions 2, 3, 5, 13, 14, 26 and 33 in the first, second and third internal repeats are shown against a white background. POSITION: numbering within one repeat; SEQUENCE: numbering throughout the sequence.

Results III

Structure of a Full-consensus Designed Ankyrin

the C-terminal repeats. The RMSD values for Cα atoms of residues 1–139 for the pairs NI₃C/E3_5, NI₃C/E3_19 and E3_5/E3_19 are 0.29 Å, 0.39 Å, and 0.36 Å, respectively. Shape complementarity (SC) values were analyzed between internal repeats and fragments thereof (see Materials and Methods). They were, in most cases, higher between internal consensus repeats than between the capping repeats and the adjacent internal consensus repeats (Table 1). This also illustrates the success of consensus design, as the internal repeats can apparently be very well stacked.

Hydrogen bond network

The total numbers of intramolecular hydrogen bonds (HB) in NI₃C, E3_5 and E3_19 are 152, 149 and 152, respectively, excluding those involved in binding sulfate ions. Interrepeat HB are least frequent in NI₃C, with 14 HB, compared to 16 and 17 HB in E3_5 and E3_19, respectively. Due to the high sequence identity, the hydrogen-bonding pattern is very similar in all three structures, as indicated by 119 common HB. The main-chain hydrogen-bonding pattern of the conserved TPLH sequence motif (residues 6–9) at the beginning of the first α-helix of every repeat is identical in NI₃C, E3_5 and E3_19. Small structural differences between different types of amino acids at the randomized positions enable the formation of specific HB. In NI₃C, E3_5 and E3_19, there are 22, 17 and 19 specific HB, respectively.

A detailed analysis of the total number of HB at the randomized positions reveals that 10 out of 34 HB in NI₃C, 5 out of 30 HB in E3_5 and 6 out of 30 HB in E3_19 are formed by side chains, again excluding those involved in binding sulfate ions. In contrast, the majority of the randomized residues contribute with their backbone atoms to the overall hydrogen-bonding network.

The rather modest number of side-chain HB from randomized residues in E3_5 and E3_19 shows that these residues are more relevant to the binding of the target molecules rather than to the formation of extended hydrogen-bonding networks on the surface of the molecules (Table 1). The side chains at the previously randomized positions specified in the full-consensus NI₃C molecule are involved in highly regular charge–charge interactions. This charge network is extended by sulfate ions from the crystallization buffer by bridging four arginine and two tyrosines residues in the first two repeats (Fig. 1b).

Charge network

Previous reports on DARPins[3,4,16] emphasized the highly regular hydrogen-bonding patterns between "framework" parts (i.e., contributed by the constant part of the sequence as important for stability). The designed ankyrins displayed a significantly higher thermodynamic stability compared to naturally occurring AR proteins,[3,4] which do not have such regular hydrogen-bonding networks and show greater variability.

The full-consensus AR NI₃C possesses, in addition, a highly regular charge distribution, which spans about half of the protein surface and causes an even higher thermodynamic stability compared to other library members. NI₃C contains a regular array of lysine and aspartic acid residues at positions 2 and 3 in the β-turns of internal repeats, as well as arginine and glutamic acid residues at positions 13 and 14 at the C-terminal ends of helices in the concave binding region. Arg23 positioned in the N-terminal capping repeat and Lys144 in the β-turn of the C-terminal capping repeat extend the charge network. The electron density of Arg23 suggests a double conformation. Both conformations allow the formation of salt bridges with a first sulfate ion. In the concave binding region, three equally spaced sulfate ions form salt bridges with Arg56, Arg89 and Arg122 (corresponding to position 13 of the internal repeat) and HB with Tyr48 and Tyr81 (position 5 in the internal repeat). In addition, these arginine residues participate in salt bridges with Glu57, Glu90 and Glu123 (corresponding to position 14 of the internal repeats) (Fig. 1b).

Thermal denaturation measurements

The thermal denaturation of NI₃C was monitored at 222 nm using CD spectroscopy. Under physiological conditions, NI₃C, unlike E3_5 and E3_19,[3,17] did not show a transition in thermal denaturation experiments.[14] In order to denature the protein

Table 1. Summary of the structural analysis between NI₃C, E3_5 and E3_19

Property	Ankyrins	NI₃C	E3_5	E3_19	
Identity (%)	NI₃C	100	89	88	
	E3_5	88	100	88	
	E3_19	87	87	100	
SC (shape complementary)	NI₃C	0.714 (SC$_{N-1}$)	0.839 (SC$_{1-2}$)	0.803 (SC$_{2-3}$)	0.705 (SC$_{3-C}$)
	E3_5	0.722 (SC$_{N-1}$)	0.762 (SC$_{1-2}$)	0.805 (SC$_{2-3}$)	0.798 (SC$_{3-C}$)
	E3_19	0.702 (SC$_{N-1}$)	0.812 (SC$_{1-2}$)	0.769 (SC$_{2-3}$)	0.739 (SC$_{3-C}$)
HB (overall)	Total	152	149	152	
	Unique	22	17	19	
	Common				119
HB (randomized residues)	Total	34	30	30	
	Side chain	10	5	6	
	Main chain	24	25	24	

completely and to reach the posttransition baseline, thermal denaturation measurements were performed in the presence of 4 M guanidinium hydrochloride Gdn·HCl. Denaturation experiments of NI₃C in Gdn·HCl had revealed a complex unfolding mechanism, which was interpreted as a partial unfolding of the C-terminal capping repeat prior to the main transition.[13,14]

A remarkable shift towards higher transition temperatures in the thermal melting curves of NI₃C was found with increasing sulfate concentrations (Fig. 3a and b). The increase in melting temperature is much more modest with increasing sodium chloride concentrations. The analysis of T_m as a function of ionic strength revealed that T_m increases linearly with ionic strength, which is in agreement with previous studies,[18] but the slope of the regression line is significantly higher for sulfate (0.0092 °C mM^{-1}; R^2=0.9884) than for phosphate (0.0064 °C mM^{-1}; R^2=0.9767) or chloride (0.0021 °C mM^{-1}; R^2=0.9268) ions (Fig. 3c). In contrast, E3_19, which does not have specific sulfate-binding sites and whose limited stability has been proposed to be a consequence of local repulsions,[17] is also stabilized by sodium chloride and sulfate (Fig. 3d). In this case, the melting point depends only on ionic strength and is independent of salt type.

Discussion

Thermal and chemical stabilities are highly relevant to the biotechnological and biomedical applications of proteins and are thus an important design goal. Consensus design exploits sequence statistics and is based on the assumption that the most frequently occurring residues in a first approximation are correlated with molecules of high thermal stability.[19] As most random mutations are destabilizing, only those protein variants that have stabilizing residues elsewhere can tolerate them. The consensus of all naturally occurring sequences would be expected to reflect these favorable residue combinations. A structural inspection is still required to avoid designing mutually exclusive residues. Compared to naturally occurring ankyrin proteins with similar numbers of repeats, consensus-designed AR proteins showed higher thermodynamic stabilities and increased stabilities towards chemical denaturants.[5,16]

The previous consensus design of DARPins has been limited to "framework" positions. The original goal was to create a library with the potential to recognize a wide range of target molecules. Consequently, residues at the randomized positions were allowed to vary and exerted attractive and repulsive charge interactions. Nonetheless, key residues specifying intrarepeat and interrepeat HB were present in every one of the internal DARPin repeats, but not necessarily in natural AR proteins, thereby partially explaining the greater stability of the designed library molecules.[3,16]

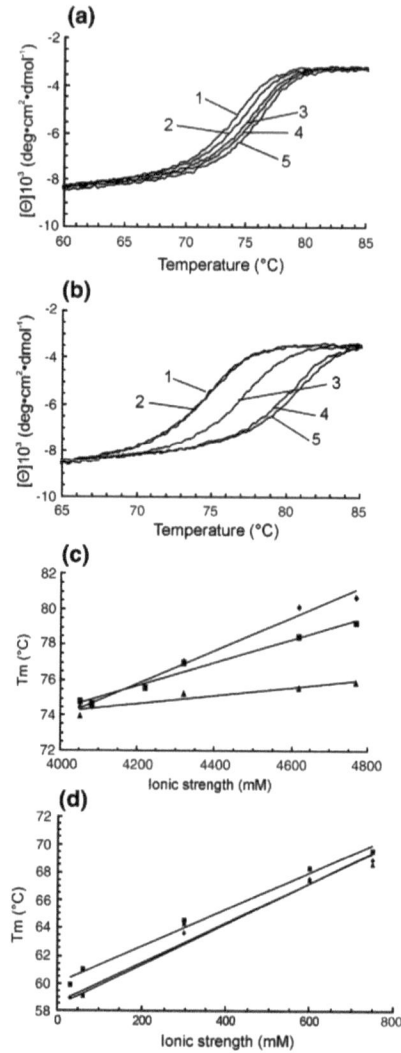

Fig. 3. Thermal melting curves of NI₃C in 20 mM Hepes, 4 M Gdn·HCl (pH 7.4) and different salt conditions: (a) sodium chloride: 30 mM (curve 1), 60 mM (curve 2), 300 mM (curve 3), 600 mM (curve 4) and 750 mM (curve 5); (b) sodium sulfate: 10 mM (curve 1), 20 mM (curve 2), 100 mM (curve 3), 200 mM (curve 4) and 250 mM (curve 5). (c) Melting points (T_m) of NI₃C plotted as a function of ionic strength for different sodium chloride (▲), sodium sulfate (♦) and sodium phosphate (■) concentrations. (d) Melting points of E3_19 as a function of ionic strengths in the same buffers as for NI₃C, but omitting Gdn·HCl. The estimated errors on T_m are ±0.5 °C.

Results III

Structure of a Full-consensus Designed Ankyrin

In the present work, we investigated the structural consequences of a full-consensus design where the sequence variability of internal repeats has been eliminated. In order to correlate structural features with the increased thermal stability of the full-consensus protein, we compared the crystal structures of the full-consensus ankyrin NI$_3$C with the previously published consensus ankyrins E3_5[16] and E3_19[17] (PDB codes 1MJO and 2BKG, respectively). A comparison of RMSD values (Table 1) and visual inspection of Cα–Cα distance plots (data not shown) indicate no major structural differences between the molecules, except within the last repeat. They have almost the same number of HB, and most of them are common between all three molecules. Contacts between repeat interfaces were measured qualitatively by analyzing shape complementarities.[20] Differences in SC values between internal repeats of NI$_3$C, E3_5 and E3_19 indicate that the stackings of internal repeats differ due to minor side-chain rearrangements. It is unlikely, however, that these differences explain the higher stability of NI$_3$C.

Charges and charge networks

The major difference between NI$_3$C, E3_5 and E3_19 is observed in the surface charge distribution. As explained above, the NI$_3$C design was based on sequence statistics, leading to a charge network that must have been selected during the evolution of AR proteins. One may speculate that residues involved in stabilizing charge networks are most strongly selected and thus dominate sequence statistics of surface residues, whereas residues that are required for target recognition cancel out in the alignments of whole sequence families.

In the work of Mosavi et al.[4], a different full-consensus sequence had been derived. In their work, positions 3 and 14 were chosen as uncharged (both asparagine), and an additional charge was introduced by an arginine at position 5 (tyrosine in this work). Additionally, there were two charge reversals in the framework of helix 2, which are lysine at position 21 (here glutamic acid because of its occurrence in GA-binding protein β1) and glutamic acid at position 25 (here lysine because of the negative charge at position 21). In the design of Mosavi et al.[4] (PDB entries 1N0Q and 1N0R), these residues generated a cluster of positively charged residues on the concave side of the molecule that might be electrostatically unfavorable (Fig. 4a). In contrast to this, the surface potential of the NI$_3$C molecule is more balanced in this area (Fig. 4b), especially when considering the protein without the sulfate ions.

Nature has used a multitude of strategies for the adaptation of proteins to life at high temperatures.[22] Surface charges were, for quite some time, considered as rather unimportant for protein stability. It was argued that the high dielectric constant of the solvent would decrease the strength of charge–charge interactions.[23] Recently, it was shown that

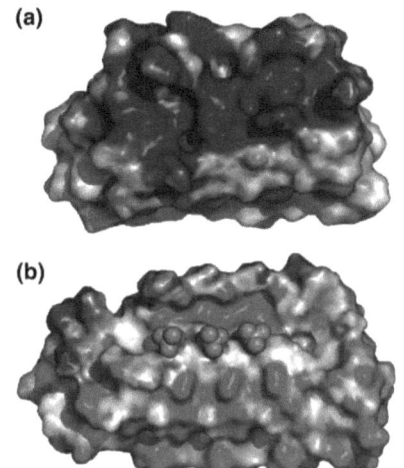

Fig. 4. Electrostatic surface potentials, calculated using the Adaptive Poisson–Boltzmann Solver[21] of (a) 4ANK (PDB code 1N0R) and (b) NI$_3$C. Both representations show the concave binding site with the N- and C-termini on the right and left sides, respectively. Bound sulfate ions in the NI$_3$C molecule are shown as spheres. Both surfaces were in blue and red for positive and negative electrostatic charge densities (same scale), respectively.

computational redesign of surface charges can improve protein stability significantly,[24] and structural genomics data on proteins from *Thermotoga maritima* revealed a significant increase in the density of salt bridges in proteins from this organism compared to their mesophilic counterparts.[25] In contrast, the importance of oligomerization order, HB and secondary structure was found to be less pronounced than previously assumed. Similar results were found in another comprehensive structural bioinformatics study[26] and in more specialized studies on individual protein families.[27–30] The high number of salt bridges in hyperthermophilic proteins was explained by the diminishing desolvation penalty for salt bridges at increasing temperatures.[31] Therefore, the contribution of salt bridges to protein stability becomes more important at higher temperatures. NI$_3$C has the smallest and most balanced negative charge excess among all three molecules studied here and, thus, essentially fewer local repulsions. A higher charge imbalance on E3_19 compared to E3_5 was proposed to be the major cause of its decreased stability.[17,32]

Another possible explanation for the extraordinarily high temperature stability, even in the absence of sulfate ions, is the flexibility of the salt bridge network, giving it an entropic advantage. Although the surface-exposed ion pair network looks rigid in the present NI$_3$C structure, only modest side-chain rearrangements would be necessary to swap between intrarepeat and interrepeat salt bridges. We suggest that the sum of all favorable charge interactions (and the absence of unfavorable ones) of the full-consensus ankyrin NI$_3$C is responsible for its significantly increased thermal stability. It is remarkable that the sequence statistics, at least in part, mirror these favorable properties.

Sulfates in the charge network

Three regularly arranged sulfate ions form the core of the salt bridge network on the concave side of the molecule. Thermal melting experiments in the presence and in the absence of sulfate ions clearly confirm the contribution of sulfate to the thermal stability of the full-consensus NI$_3$C ankyrin. It should be noted that thermal stability is already very high in the absence of sulfate. Even though the surface potential is balanced (Fig. 4b), the protein can bind three sulfate ions, which increase the total negative charge excess of the molecule from 12 e$^-$ to 18 e$^-$. Because of the alternating arrangement of the charges, sulfate binding is energetically favored, and the stabilizing effect may be rationalized by the very short hydrogen-bonding distance of 2.4 Å between the sulfate ions and tyrosine residues at position 5 (Fig. 1b).

Summary and conclusion

The crystal structure and the thermal denaturation studies of the designed full-consensus AR protein NI$_3$C suggest that the extended salt bridge network positively influences the stability of the protein. This protein can only be denatured by heating in the presence of 4 M Gdn·HCl. NI$_3$C was designed based mainly on sequence statistics, suggesting that this charge network among previously randomized residues must also be a hidden design feature of natural ARs. In naturally occurring AR proteins, this feature is no longer visible because the sequences have drifted due to functional selection. Designing such charge interaction networks may thus be a strategy for the development of extremely stable proteins for biomedical and other biotechnical applications. However, since many of these residues normally participate in binding, a compromise between function and thermal stability by surface charge interactions will have to be found.

Materials and Methods

Expression and purification

The full-consensus NI$_3$C ankyrin was expressed in *E. coli* strain XL-1 blue in LB medium supplemented with 1% glucose, 100 µg/ml ampicillin and 15 µg/ml tetracycline. After growing to an OD$_{600}$ of approximately 0.8–1.0 at 37 °C, the cultures were induced with 0.5 mM IPTG, and growth continued for 5 h. After centrifugation and resuspension in 50 mM Tris, 500 mM NaCl and 20 mM imidazole at pH 8.0, the cells were disrupted using a French press. The protein was purified in self-packed columns containing a Ni-NTA matrix (Qiagen) in accordance with the manufacturer's instructions. The protein was further purified by gel filtration on a HiLoad 26/60 Superdex 75 (Pharmacia Biotech) with an ÄKTA Prime system (Amersham, Pharmacia Biotech) in 10 mM Hepes and 10 mM NaCl (pH 7.4) and concentrated to 20 mg/ml using a Centricon (Millipore USA) with 3-kDa molecular mass cutoff.

Crystallization

Crystals for X-ray diffraction data collection were grown in 24-well crystallization plates using the hanging-drop vapor-diffusion method. A 1-µl protein solution was mixed with 1 µl of reservoir solution containing 2.7 M (NH$_4$)$_2$SO$_4$ and 100 mM Tris (pH 8.5). The crystals grown under these conditions were soaked in reservoir solution supplemented with 20% glycerol as cryoprotectant and flash-cooled in a nitrogen stream at 100 K. The X-ray diffraction data of one single crystal were collected using CuKα radiation generated by a Nonius FR591 rotating anode generator.

Structure solution and refinement

Diffraction data up to 2.05 Å resolution on a total of 180 frames were recorded within an oscillation range of 1°. The data were processed with program XDS.[33] A Matthews coefficient of $V_M = 2.4$ Å^3 Da^{-1} suggested one molecule in the asymmetric unit. Initial phases were obtained by molecular replacement using the program

Table 2. Statistics for data collection and refinement

Data collection	
Space group	P6$_1$
Cell dimensions	$a = 74.48$ Å, $b = 74.48$ Å, $c = 50.99$ Å,
	$\alpha = \beta = 90°$, $\gamma = 120°$
Resolution limits (Å)	64.5–2.05
Observed reflections	78,634 (overall); 10,232 (unique)
Completeness (%)	99.8 (98.2)
Redundancy	7.7 (7.1)
R_{sym} (% on I)	7.3 (37.0)
Refinement	
Resolution range (Å)	64.5–2.05
R-factor/R_{free} (%)	18.6/22.6
Ordered water molecules	110
RMSD from ideal geometry	
Bond length (Å)	0.010
Bond angles (°)	1.109
Average B-factor (Å^2)	26.3
Residues in the most favored region (%)	91.9
Residues in the additionally allowed region (%)	8.1
Residues in the generously allowed region (%)	0.0
Residues in the disallowed region (%)	0.0

Numbers inside parentheses refer to the highest resolution shell (2.1–2.05 Å).

Results III

AMoRe,[34] with the structure of E3_5 (PDB code 1MJO) as a search model. The structure was refined to final R-factors of $R_{work} = 18.6\%$ and $R_{free} = 22.6\%$ in the space group $P6_1$ using the program REFMAC5.[35] Model building was performed with program O.[36] Statistics for diffraction data and structure refinement are summarized in Table 2. The program PROCHECK[37] was used to evaluate deviations from standard geometries, and the programs HBPLUS[38] and SC[20] were used to analyze the hydrogen-bonding pattern and calculations of surface complementarities. Although the SC program was originally developed to calculate the surface complementarity between separated subunits, we used this method to investigate intramolecular shape complementarities. For this purpose, the molecule was split into fragments by inserting artificial chain breaks between residues 31/32, 64/65, 97/98 and 130/131. The SC_{N-1}, SC_{1-2}, SC_{2-3} and SC_{3-C} values refer to the interface SC values of neighboring repeats. Figures were generated with PyMOL.[39]

Thermal denaturation experiments

Thermal denaturation experiments were performed with a Jasco J-715 instrument (Jasco, Japan). CD data were recorded at a protein concentration of 40 μM in the temperature range between 5 °C and 95 °C within 90 min at a wavelength of 222 nm (measuring interval, 10 s; bandwidth, 2 nm). The buffers contained 20 mM Hepes, 4 M Gdn·HCl (pH 7.4) and salt concentrations of between 30 and 750 mM sodium chloride or of between 10 and 250 mM sodium sulfate. The same buffers without Gdn·HCl were used for heat denaturation experiments with E3_19. Ionic strengths were calculated as follows: $I = 1/2 \sum c_i z_i^2$, where c = concentration and z = charge of ion.

Protein Data Bank accession number

Coordinates and diffraction data have been deposited at the PDB under accession number 2QYJ.

Acknowledgement

Financial support from the Swiss National Science Foundation and the Baugartenstiftung (Zurich, Switzerland) is gratefully acknowledged.

References

1. Andrade, M. A., Perez-Iratxeta, C. & Ponting, C. P. (2001). Protein repeats: structures, functions, and evolution. *J. Struct. Biol.* **134**, 117–131.
2. Bork, P. (1993). Hundreds of ankyrin-like repeats in functionally diverse proteins: mobile modules that cross phyla horizontally? *Proteins: Struct. Funct. Genet.* **17**, 363–374.
3. Binz, H. K., Stumpp, M. T., Forrer, P., Amstutz, P. & Plückthun, A. (2003). Designing repeat proteins: well-expressed, soluble and stable proteins from combinatorial libraries of consensus ankyrin repeat proteins. *J. Mol. Biol.* **332**, 489–503.
4. Mosavi, L. K., Minor, D. L., Jr & Peng, Z. Y. (2002). Consensus-derived structural determinants of the ankyrin repeat motif. *Proc. Natl Acad. Sci. USA*, **99**, 16029–16034.
5. Mosavi, L. K. & Peng, Z. Y. (2003). Structure-based substitutions for increased solubility of a designed protein. *Protein Eng.* **16**, 739–745.
6. Stumpp, M. T., Forrer, P., Binz, H. K. & Plückthun, A. (2003). Designing repeat proteins: modular leucine-rich repeat protein libraries based on the mammalian ribonuclease inhibitor family. *J. Mol. Biol.* **332**, 471–487.
7. Sedgwick, S. G. & Smerdon, S. J. (1999). The ankyrin repeat: a diversity of interactions on a common structural framework. *Trends Biochem. Sci.* **24**, 311–316.
8. Amstutz, P., Binz, H. K., Parizek, P., Stumpp, M. T., Kohl, A., Grütter, M. G. *et al.* (2005). Intracellular kinase inhibitors selected from combinatorial libraries of designed ankyrin repeat proteins. *J. Biol. Chem.* **280**, 24715–24722.
9. Binz, H. K., Amstutz, P., Kohl, A., Stumpp, M. T., Briand, C., Forrer, P. *et al.* (2004). High-affinity binders selected from designed ankyrin repeat protein libraries. *Nat. Biotechnol.* **22**, 575–582.
10. Kohl, A., Amstutz, P., Parizek, P., Binz, H. K., Briand, C., Capitani, G. *et al.* (2005). Allosteric inhibition of aminoglycoside phosphotransferase by a designed ankyrin repeat protein. *Structure*, **13**(8), 1131–1141.
11. Schweizer, A., Roschitzki-Voser, H., Amstutz, P., Briand, C., Gulotti-Georgieva, M., Prenosil, E. *et al.* (2007). Inhibition of Caspase-2 by a designed ankyrin repeat protein: specificity, structure and inhibition mechanism. *Structure*, **15**(15), 625–635.
12. Sennhauser, G., Amstutz, P., Briand, C., Storchenegger, O. & Grütter, M. G. (2007). Drug export pathway of multidrug exporter AcrB revealed by DARPin inhibitors. *PLoS Biol.* **5**(1), e7.
13. Interlandi, G., Wetzel, S.K., Settanni, G., Plückthun, A., Caflisch, A. (2007). Molecular dynamics simulations and chemical denaturation experiments. *J. Mol. Biol.* doi:10.1016/j.jmb.2007.09.042.
14. Wetzel, S. K., Settanni, G., Kenig, M., Binz, H. K. & Plückthun, A. (2008). Folding and unfolding mechanism of highly stable full consensus ankyrin repeat proteins. *J. Mol. Biol.* **376**, 241–257.
15. Batchelor, A. H., Piper, D. E., de la Brousse, F. C., McKnight, S. L. & Wolberger, C. (1998). The structure of GABPalpha/beta: an ETS domain-ankyrin repeat heterodimer bound to DNA. *Science*, **279**, 1037–1041.
16. Kohl, A., Binz, H. K., Forrer, P., Stumpp, M. T., Plückthun, A. & Grütter, M. G. (2003). Designed to be stable: crystal structure of a consensus ankyrin repeat protein. *Proc. Natl Acad. Sci. USA*, **100**, 1700–1705.
17. Binz, H. K., Kohl, A., Plückthun, A. & Grütter, M. G. (2006). Crystal structure of a consensus-designed ankyrin repeat protein: implications for stability. *Proteins: Struct. Funct. Genet.* **65**, 280–284.
18. Tadeo, X., Pons, M. & Millet, O. (2007). Influence of the Hofmeister anions on protein stability as studied by thermal denaturation and chemical shift perturbation. *Biochemistry*, **46**, 917–923.
19. Forrer, P., Binz, H. K., Stumpp, M. T. & Plückthun, A. (2004). Consensus design of repeat proteins. *ChemBioChem*, **5**, 183–189.
20. Lawrence, M. C. & Colman, P. M. (1993). Shape complementarity at protein/protein interfaces. *J. Mol. Biol.* **234**, 946–950.

21. Baker, N. A., Sept, D., Joseph, S., Holst, M. J. & McCammon, J. A. (2001). Electrostatics of nanosystems: application to microtubules and the ribosome. *Proc. Natl Acad. Sci. USA*, **98**, 10037–10041.
22. Jaenicke, R. & Böhm, G. (1998). The stability of proteins in extreme environments. *Curr. Opin. Struct. Biol.* **8**, 738–748.
23. Schueler-Furman, O., Wang, C., Bradley, P., Misura, K. & Baker, D. (2005). Progress in modeling of protein structures and interactions. *Science*, **310**, 638–642.
24. Strickler, S. S., Gribenko, A. V., Gribenko, A. V., Keiffer, T. R., Tomlinson, J., Reihle, T. *et al.* (2006). Protein stability and surface electrostatics: a charged relationship. *Biochemistry*, **45**, 2761–2766.
25. Robinson-Rechavi, M., Alibes, A. & Godzik, A. (2006). Contribution of electrostatic interactions, compactness and quaternary structure to protein thermostability: lessons from structural genomics of *Thermotoga maritima*. *J. Mol. Biol.* **356**, 547–557.
26. Alsop, E., Silver, M. & Livesay, D. R. (2003). Optimized electrostatic surfaces parallel increased thermostability: a structural bioinformatic analysis. *Protein Eng.* **16**, 871–874.
27. Cheung, Y. Y., Lam, S. Y., Chu, W. K., Allen, M. D., Bycroft, M. & Wong, K. B. (2005). Crystal structure of a hyperthermophilic archaeal acylphosphatase from *Pyrococcus horikoshii*—structural insights into enzymatic catalysis, thermostability, and dimerization. *Biochemistry*, **44**, 4601–4611.
28. Corazza, A., Rosano, C., Pagano, K., Alverdi, V., Esposito, G., Capanni, C. *et al.* (2006). Structure, conformational stability, and enzymatic properties of acylphosphatase from the hyperthermophile *Sulfolobus solfataricus*. *Proteins: Struct. Funct. Genet.* **62**, 64–79.
29. Karlstrom, M., Steen, I. H., Madern, D., Fedoy, A. E., Birkeland, N. K. & Ladenstein, R. (2006). The crystal structure of a hyperthermostable subfamily II isocitrate dehydrogenase from *Thermotoga maritima*. *FEBS J.* **273**, 2851–2868.
30. Scandurra, R., Consalvi, V., Chiaraluce, R., Politi, L. & Engel, P. C. (1998). Protein thermostability in extremophiles. *Biochimie*, **80**, 933–941.
31. Elcock, A. H. (1998). The stability of salt bridges at high temperatures: implications for hyperthermophilic proteins. *J. Mol. Biol.* **284**, 489–502.
32. Yu, H., Kohl, A., Binz, H. K., Plückthun, A., Grütter, M. G. & van Gunsteren, W. F. (2006). Molecular dynamics study of the stabilities of consensus designed ankyrin repeat proteins. *Proteins: Struct. Funct. Genet.* **65**, 285–295.
33. Kabsch, W. (1993). Automatic processing of rotation diffraction data from crystals of initially unknown symmetry and cell constants. *J. Appl. Crystallogr.* **26**, 795–800.
34. Navaza, J., Panepucci, E. H. & Martin, C. (1998). On the use of strong Patterson function signals in many-body molecular replacement. *Acta Crystallogr. Sect. D*, **54**, 817–821.
35. Murshudov, G. N., Vagin, A. A., Lebedev, A., Wilson, K. S. & Dodson, E. J. (1999). Efficient anisotropic refinement of macromolecular structures using FFT. *Acta Crystallogr. Sect. D*, **55**, 247–255.
36. Jones, T. A., Zou, J. Y., Cowan, S. W. & Kjeldgaard, M. (1991). Improved methods for building protein models in electron density maps and the location of errors in these models. *Acta Crystallogr. Sect. A*, **47**, 110–119.
37. Laskowski, R. A., Moss, D. S. & Thornton, J. M. (1993). Main-chain bond lengths and bond angles in protein structures. *J. Mol. Biol.* **231**, 1049–1067.
38. McDonald, I. K. & Thornton, J. M. (1994). Satisfying hydrogen bonding potential in proteins. *J. Mol. Biol.* **238**, 777–793.
39. DeLano, W. L. (2002). *The PyMOL Molecular Graphics System*. DeLano Scientific, Palo Alto, CA, USA.

Results **IV**

2.4 Stepwise unfolding of ankyrin repeats in a single protein revealed by atomic force microscopy (*IV*)

Here the unfolding mechanism of the longest full-consensus DARPin NI_6C is analyzed using mechanical pulling as unfolding force. The results suggest a sequential mechanism, where the repeats unfold individually. AR proteins are found in many different cellular functions, such as cytoskeletal organization, where mechanical stress and deformations may be involved. One example is a spectrin binding AR protein.

In a similar study using the 12-repeat protein ankyrin-R it was shown that AR exhibit tertiary-structure-based elasticity and behave as a linear and fully reversible spring in single-molecule measurements by atomic force microscopy (AFM). These results have implications in the field of mechanotransduction, i.e. modulating activity of ankyrin-associated transporters in response to mechanical strain.[30]

Li, L. W., Wetzel, S., Plückthun, A. & Fernandez, J. M. (2006). Stepwise unfolding of ankyrin repeats in a single protein revealed by atomic force microscopy. *Biophys. J.* **90**, L30-L32.

IV

Results **IV**

Biophysical Journal: Biophysical Letters

Stepwise Unfolding of Ankyrin Repeats in a Single Protein Revealed by Atomic Force Microscopy

Lewyn Li,* Svava Wetzel,[†] Andreas Plückthun,[†] and Julio M. Fernandez*
*Department of Biological Sciences, Columbia University, New York, New York 10027; and [†]Biochemisches Institut, Universität Zürich, CH-8057 Zürich, Switzerland

ABSTRACT Using single-molecule atomic force microscopy, we find that a protein consisting of six identical ankyrin repeat units flanked by N- and C-terminal modules (N6C) unfolds in a stepwise, unit-by-unit fashion under a mechanical force. Stretching a N6C molecule results in a sawtooth pattern fingerprint, with as many as six peaks separated by ~10 nm and an average unfolding force of 50 ± 20 pN. Our results demonstrate that a stretching force can unfold multiple repeat units individually in a single protein molecule, despite extensive hydrophobic interactions between adjacent units.

Received for publication 23 November 2005 and in final form 14 December 2005.

Address reprint requests and inquiries to Julio M. Fernandez, Tel.: 212-854-9141; Fax: 212-854-4619;
E-mail: jfernandez@columbia.edu.

Proteins containing repeats of the same structural motif are common in nature. The ankyrin (ANK) repeat is a 33-residue L-shaped motif, consisting of two antiparallel α-helices linked by a short loop (1). A β-turn forms the base of the L and connects two consecutive repeats (Fig. 1 *a*). Adjacent ANK repeats are stabilized by extensive hydrophobic and hydrogen bonding interactions (1). Recent genome searches revealed 10,000–20,000 ANK repeats among viruses, prokaryotes, and eukaryotes, usually with four to six (but up to 29) consecutive repeats in one protein (1,2). The main function of ANK repeat is to mediate protein-protein interactions. ANK repeat proteins are found in many different cellular roles, including those where mechanical stress and deformations may be involved. For example, in different cell types, ankyrins bind both to spectrin and to outer membrane proteins such as ion channels, thus connecting the membrane to the cytoskeleton (3). The spectrin-actin cytoskeleton provides mechanical support for the membrane bilayer (3). Recently, a chain of 17–29 ANK repeats in the mechanosensitive transduction channel TRPA1 has been proposed to be the gating spring that controls the opening and closing of ion channels in vertebrate hair cells (4–6).

ANK repeat proteins also represent an interesting class of models in protein folding, because their modular structure and lack of long-range interactions have led to the hypothesis that each ANK repeat in a protein could unfold and refold individually. Thus far, ensemble thermal and chemical denaturation experiments of ANK repeat proteins do not show this behavior (1,7–10). However, temperature and chemicals such as urea are global unfolding agents that affect the entire protein surface. On the other hand, the application of a stretching force in single-molecule atomic force microscopy (AFM) unravels a protein along a specific direction, and may be a physiologically relevant denaturant in some cases (11–17). In this study, we decided to test the hypothesis of individual ANK repeat unfolding by mechanically stretching a consensus ANK repeat protein, N6C, using single-molecule AFM.

N6C consists of an N-terminal cap, six identical copies of a designed consensus ANK repeat and a C-terminal cap that form a contiguous domain (S. Wetzel, K. Binz, and A. Plückthun, unpublished data; see Supplementary Material). N6C is an ideal system for single-molecule AFM because the presence of multiple copies of the same ANK repeat eliminates complications from the heterogeneous mixtures of ANK repeats in natural proteins (11). The N6C protein was adsorbed onto a gold-coated coverslip and randomly picked up and stretched by an AFM tip. Stretching N6C generates a uniform sawtooth pattern in the force-extension curve with as many as six peaks regularly separated by ~10 nm (Fig. 1 *b*). Using the worm-like chain model (18), we determined that successive peaks correspond to an increase in polymer contour length (ΔL_c) of 11.5 ± 0.7 nm (Fig. 1 *d*). Since an amino acid contributes 0.38 nm in contour length and each folded ANK repeat possesses a length of 0.8 nm, the unfolding of an ANK repeat is expected to increase the contour length by (33 × 0.38–0.8 nm) = 11.5 nm. The excellent agreement between the predicted and observed change in contour length and the regular sawtooth pattern are strong indications that ANK repeats in a N6C molecule unravel individually. The average unfolding force is 50 ± 20 pN (Fig. 1 *c*). Computer simulations have shown similar piecewise unfolding of ANK repeats in a four-repeat model (6). An ANK repeat protein is different from polyproteins such as titin, tenascin, fibronectin, and ubiquitin, because all the ANK repeats belong to a single protein domain and there are extensive hydrophobic interactions and hydrogen bonds

© 2006 by the Biophysical Society
doi: 10.1529/biophysj.105.078436

Results IV

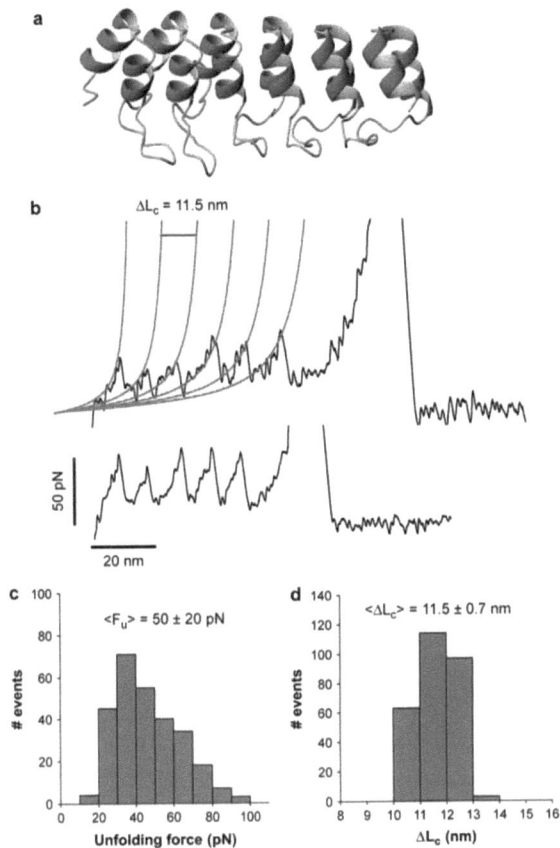

FIGURE 1 Mechanical unfolding of the ankyrin repeats in N6C with single-molecule AFM. (*a*) An ANK repeat protein structure containing six ANK repeats. This structure consists of the second to seventh repeats in human ankyrinR (19). (*b*) Representative force-extension curves for N6C. The AFM tip picked up the protein at random locations along the peptide backbone, resulting in a different number of peaks in each trace. The contour length increment, $\Delta L_c = 11.5$ nm, is only observed at high forces. (*c*) The distribution of peak forces, with $N = 277$. (*d*) The distribution of ΔL_c, with $N = 277$.

between adjacent ANK repeats. In polyproteins, each protein domain is individually folded and spatially separated, and there may only be a small number of domain-domain contacts because of the short interdomain linker (two to four amino acids) and the relative rigidity of folded protein domains.

Here we demonstrate that, under a stretching force, the ANK repeats in a protein can be unraveled one at a time. In contrast, most bulk thermal and chemical unfolding studies of ANK repeat proteins, including N6C (S. Wetzel, K. Binz, and A. Plückthun, unpublished data), show highly cooperative transitions, with zero or one unfolding or folding intermediate (1,7–10). Why do the ANK repeats in N6C behave so differently between mechanical and chemical unfolding? In mechanical unfolding, the vectorial nature of the applied force compels the protein to unfold along a well-defined direction, whereas in chemical denaturation, the entire protein surface is exposed to unfolding agents. Therefore, the reaction coordinate is probably different between mechanical and chemical denaturation, resulting in different unfolding pathways, transition states, and unfolded states. For example, the mechanical stabilities of fibronectin domains do not follow their melting temperatures (14), and the mechanical stability of a protein depends on the direction of the applied force vector (12,13). The fact that we observed individual ANK repeat unfolding under a stretching force is an indication that the direction in which a protein moves on the unfolding free energy landscape is crucial to how a protein unfolds. Under a stretching force, the ANK repeats in N6C are better described as a linear sequence of unfolding units, with little communication or cooperation among various ANK repeats.

Some ANK repeat proteins may play mechanical roles in vivo. For example, a chain of 17–29 ANK repeats in the mechanosensitive transduction channel TRPA1 has been proposed to form a curved ANK superhelix, which is a possible candidate for the gating spring that transmits forces from sound waves or motion to ion channels (4–6). Molecular dynamics simulations postulated that chains of 12–24 ANK repeats respond to a stretching force in two phases (6). In the first phase, the ANK superhelix stretches and relaxes reversibly by ~11 nm without unfolding. During the second phase, individual ANK repeats begin to unfold, representing an attractive (albeit unconfirmed) safety mechanism against extreme stimuli. The piecewise unfolding of multiple ANK repeats in N6C (Fig. 1 b) show that individual ANK repeat can indeed be peeled off by mechanical stress. Therefore, our results support the idea that, in some biological systems, ANK repeats could behave like multiple buffers linked in series; to resist damagingly high forces, ANK repeats can be sacrificed and extended one at a time, without the whole protein losing its tertiary structure. Evolution may have selected the modular design of some ANK repeat proteins for mechanical reasons.

SUPPLEMENTARY MATERIAL

An online supplement to this article can be found by visiting BJ Online at http://www.biophysj.org.

ACKNOWLEDGMENTS

This research was supported by National Institutes of Health grants No. R01 HL66030 and R01 HL61228 to J.M.F. L.L. is a Damon Runyon Fellow (DRG No. 1792-03).

REFERENCES and FOOTNOTES

1. Mosavi, L. K., T. J. Cammett, D. C. Desrosiers, and Z.-Y. Peng. 2004. The ankyrin repeat as molecular architecture for protein recognition. *Protein Sci.* 13:1435–1448.
2. Bork, P. 1993. Hundreds of ankyrin-like repeats in functionally diverse proteins: mobile modules that cross phyla horizontally? *Proteins.* 17:363–374.
3. Bennett, V., and A. J. Baines. 2001. Spectrin and ankyrin-based pathways: metazoan inventions for integrating cells into tissues. *Physiol. Rev.* 81:1353–1392.
4. Howard, J., and S. Bechstedt. 2004. Hypothesis: a helix of ankyrin repeats of the NOMPC-TRP ion channel is the gating spring of mechanoreception. *Curr. Biol.* 14:R224–R226.
5. Corey, D. P., J. Garcia-Anoveros, J. R. Holt, K. Y. Kwan, S. Y. Lin, M. A. Vollrath, A. Amalfitano, E. L. Cheung, B. H. Derfler, A. Duggan, G. S. Geleoc, P. A. Gray, et al. 2004. TRPA1 is a candidate for the mechanosensitive transduction channel of vertebrate hair cells. *Nature (Lond.).* 432:723–730.
6. Sotomayor, M., D. P. Corey, and K. Schulten. 2005. In search of the hair-cell gating-spring: elastic properties of ankyrin and cadherin repeats. *Structure.* 13:669–682.
7. Zweifel, M. E., and D. Barrick. 2001. Studies of the ankyrin repeats of the Drosophila melanogaster Notch receptor. 2. Solution stability and cooperativity of unfolding. *Biochemistry.* 40:14357–14367.
8. Binz, H. K., M. T. Stumpp, P. Forrer, P. Amstutz, and A. Plückthun. 2003. Designing repeat proteins: well-expressed, soluble and stable proteins from combinatorial libraries of consensus ankyrin repeat proteins. *J. Mol. Biol.* 332:489–503.
9. Tang, K. S., A. R. Fersht, and L. S. Itzhaki. 2003. Sequential unfolding ankyrin repeats in tumor suppressor p16. *Structure.* 11:67–73.
10. Zeeb, M., H. Rosner, W. Zeslawski, D. Canet, T. A. Holak, and J. Balbach. 2002. Protein folding and stability of human CDK inhibitor p19^{INK4d}. *J. Mol. Biol.* 315:447–457.
11. Carrion-Vazquez, M., A. F. Oberhauser, S. B. Fowler, P. E. Marszalek, S. E. Broedel, J. Clarke, and J. M. Fernandez. 1999. Mechanical and chemical unfolding of a single protein: a comparison. *Proc. Natl. Acad. Sci. USA.* 96:3694–3699.
12. Carrion-Vazquez, M., H. Li, H. Lu, P. E. Marszalek, A. F. Oberhauser, and J. F. Fernandez. 2003. The mechanical stability of ubiquitin is linkage dependent. *Nat. Struct. Biol.* 10:738–743.
13. Brockwell, D. J., E. Paci, R. C. Zinober, G. S. Beddard, P. D. Olmsted, D. A. Smith, R. N. Perham, and S. E. Radford. 2003. Pulling geometry defines the mechanical resistance of a β-sheet protein. *Nat. Struct. Biol.* 10:731–737.
14. Oberhauser, A. F., C. Badilla-Fernandez, M. Carrion-Vazquez, and J. M. Fernandez. 2002. The mechanical hierarchies of fibronectin observed with single-molecule AFM. *J. Mol. Biol.* 319:433–447.
15. Li, H., H. H.-L. Huang, C. L. Badilla, and J. M. Fernandez. 2005. Mechanical unfolding intermediates observed by single-molecule force spectroscopy in a fibronectin type III module. *J. Mol. Biol.* 345:817–826.
16. Forman, J. R., S. Qamar, E. Paci, R. N. Sandford, and J. Clarke. 2005. The remarkable mechanical strength of polycystin-1 supports a direct role in mechanotransduction. *J. Mol. Biol.* 349:861–871.
17. Schwaiger, I., A. Kardinal, M. Schleicher, A. A. Noegel, and M. Rief. 2004. A mechanical unfolding intermediate in an actin-crossing protein. *Nat. Struct. Biol.* 11:81–85.
18. Marko, J. F., and E. D. Siggia. 1995. Stretching DNA. *Macromolecules.* 28:8759–8770.
19. Michaely, P., D. R. Tomchick, M. Machius, and R. G. Anderson. 2002. Crystal structure of a 12 ANK repeat stack from human ankyrinR. *EMBO J.* 21:6387–6396.

3. Protocols
3.1 Equilibrium unfolding and refolding curve

For the preparation of an unfolding curve, two buffer solutions have to be prepared: the native buffer for the protein and a stock solution of the denaturant (GdnHCl) dissolved in the same native buffer at the same pH. In order to determine the exact concentration of GdnHCl, the refractive index (N) of each solution is measured and the molar concentration of GdnHCl calculated using the formula $57.147 * \Delta N + 38.28\, (\Delta N)^2 - 91.6\, (\Delta N)^3$, where ΔN = N(GdnHCl-PBS solution) – N(PBS). The buffers have to be filtered through 0.22 µm filters.

For a first measurement, 15 samples are usually sufficient: 5 points in the pre-transition region, 5 in the transition region and 5 points in the post-transition region. A pipetting scheme is prepared in Excel with the desired final concentrations of denaturant. The denaturant stock solution is mixed with the native buffer in different ratios to obtain 15 buffer solutions. Then, to each buffer, a constant volume of highly concentrated protein solution is added and mixed by pipetting. For better accuracy the protein volume should be around 10 to 30 µl. The final protein concentration has to be high enough to provide a good CD or fluorescence signal. In the case of the DARPins, 5-10 µM was necessary. For each denaturant concentration a buffer sample and a protein sample have to be prepared. The protein is equilibrated in the denaturant at the desired temperature that is used for the measurement.

The samples are then measured in a CD spectrometer or fluorimeter. The measurement settings used for the DARPins are as follows:

1) CD (Jasco J-715): standard sensitivity 100 mdeg, wavelength 222 nm, data pinch 5 nm, scanning mode continuous, response 4 sec, band width 2 nm, 2 min measurement time.
2) Fluorescence (PTI Alpha Scan spectrofluorimeter): excitation λ_1 and emission spectrum λ_2 – λ_3, photomultiplier 800 V, slit size 5 nm, method – step size 1 nm, integration time 0.5 s, averages 3 – shutters automatic, acquire emission scan – excitation 248 nm, emission 263,5 nm. Check the wavelength setting always by looking at the number indicated on the monochromator (emission λ_2) and then, in case of necessity, correct the number in the corresponding window for the wavelength in the software.

The solutions are filled into the cuvette, equilibrated at the desired temperature for 1 min and then measured. The cuvette is not washed between each measurement, but it is important to start with native buffer and then to continue with increasing denaturant concentrations. For fluorescence cuvettes it is best to use a Pasteur pipet with a silicon tubing on the tip to transfer the solution into the cuvette and to remove it in one step.

Protocols

In order to make sure that the protein is at equilibrium, the measurement is repeated after 2 days. If the signal remains the same, the equilibration time was enough. Usually equilibration overnight is sufficient.

For the refolding curve the protein stock solution has to be unfolded first. The denaturant concentration at the beginning of the post-transition region of the previously measured unfolding curve is chosen for the preparation of the unfolded protein stock solution. After equilibration of one day (or the time necessary to reach equilibrium), a similar set of 15 buffer samples is prepared and a constant volume of the unfolded protein is added. The protein samples are equilibrated for another night and measured.

For the data analysis, the buffer signal has to be deduced from the protein signal and plotted against the denaturant concentration. The curves can then be fitted using the two-and three-state folding models (equations see Appendix). The condition for reversible folding is fulfilled if refolding curve and unfolding curve overlay.

In the case of fluorescence measurements, the analysis can be done in several ways: usually the maximal emission intensity is plotted versus the denaturant concentration, but also the area or the center of gravity under the emission spectrum can be calculated and plotted versus denaturant concentration.[47,48]

3.2 Unfolding and refolding kinetics measured in a stopped-flow instrument

Equivalent to the equilibrium measurement, two buffer solutions have to be prepared.

For a first chevron plot, it is enough to have 6 points for the folding limb and 6 points for the unfolding limb, i.e. 12 buffer solutions with different denaturant concentrations. For measuring unfolding kinetics, native protein is mixed with denaturant. For refolding kinetics, unfolded protein is mixed with native buffer and denaturant buffers of lower denaturant concentration. A chevron plot can be measured in one day. For the determination of the denaturant buffer concentrations, the information of the equilibrium transition curve is needed. Unfolding is started from the denaturant concentration of the transition midpoint (D_m), refolding is performed from native buffer to the denaturant concentration below D_m. Reasonable volumes for the solutions are: 10 ml protein solution, 20-30 ml buffer of each concentration.

Depending on the spectroscopic probe, the protein stock concentration has to be determined. In the case of the DARPins, which are rich in α-helix, and therefore exhibit a strong CD signal at 225 nm, a 200 µM (2 mm path) or 50 µM (10 mm path) protein stock solution was prepared, depending on which cuvette path length was used. If a fluorophor is used for fluorescence detection, the concentration of the protein can be much lower to achieve a good signal.

Protocols

The stopped-flow instrument Pi-Star from Applied Photophysics has a cylindrical cuvette, with the dimensions 2 mm (deep) x 10 mm (long) x 1 mm (wide) and 80 µl is the minimal volume that has to be pushed through the cuvette in order to measure kinetics. Indeed, during my measurements, I used always a shot volume of 250 µl, i.e. three time higher than necessary. The cuvette can be turned, so that measurements in the 2 mm path as well as in the 10 mm path are possible. If you want to save protein, the 10 mm path is recommended.

For the protein sample a 200 µl syringe and for the buffer solution a 2 ml syringe are used. Using these syringe volumes, a mixing ratio of 1:10, typical for protein folding measurements, is obtained. The schematic configuration of the stopped-flow system for single mixing experiments and double mixing experiments is represented in Fig. 12. Single mixing is used for the classical folding and unfolding measurements, while double mixing can serve for interrupted refolding or unfolding measurements as described by Schmid et al.[17]

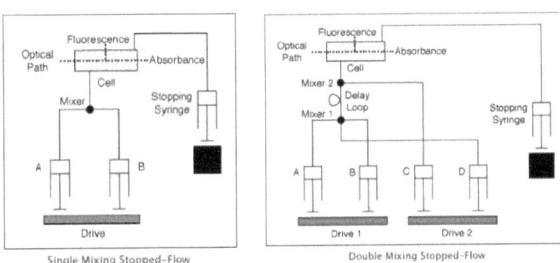

Fig. 12: Schematic representation of the stopped-flow mixing units: single mixing and double mixing. At the position of absorbance detection, also CD detection is possible.

1) Switch on the instrument and the lamp, the water bath and set the measurement temperature, then switch on the computer and wait 30 min for heating up the lamp. Open the liquid nitrogen bottle to apply a pressure of 2 bar on the instrument.

2) Enter all the settings into the Pi-Star software. In the control panel window enter: CD (or Fluorescence), AutoPM, Logarithmic data acquisition, over sampling, external trigger, monochromator λ = 225 nm (for fluorescence: the corresponding excitation wavelength); in the details sub window: enter slit size 4 nm for entrance and exit slit (CD measurements, for fluorescence to be determined depending on fluorophore).

3) Wash the syringes with water 10 times (using drive option) and measure the CD/Fluorescence signal in order to have a baseline value.

4) Fill the syringes with native buffer (flush tubing) and measure the baseline.

5) Then exchange the buffer against the native protein stock solution in the smaller syringe (200 µl volume), see Fig. 13.

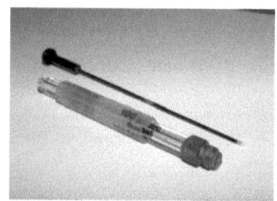

Fig. 13: Hamilton syringe used in the stopped-flow instrument π*: the 100 µl syringe can be used to obtain mixing ratios of 1:20.

6) Measure the protein baseline, meaning that native protein is mixed with native buffer. This means that the tubings have to be flushed before the measurement to make sure that really the protein is pushed through the cuvette. Usually I needed to flush up to 10 times before measurement.

7) Exchange the native buffer against denaturant buffer with increasing denaturant concentrations. The unfolding kinetics are measured for each denaturant concentration. For the change of the buffer solution, the "empty" option in the control software is used. Here the solution is not pushed through the cuvette, but shot into the waste directly. One can control the exchange of the buffer observing the stop syringe, when the buffer is completely exchanged, usually after five shots, the measurement can be started.

First, the duration of the measurement has to be decided. Single shots are carried out. When the signal reaches a plateau, the reaction is over. Then, several shots are performed with a fixed measurement time. The number of repetitions depends on the measurement time: kinetics of 1 second have to be repeated at least 14 times in order to have a good signal to noise ratio, kinetics of 20 seconds only 7 times. This is judged from the quality of the signal; the more repeats, the better the signal, but also more protein is used.

8) The buffer syringe is washed and filled with native buffer, the protein syringe is filled with the unfolded protein solution, the tubings have to be flushed.

9) The refolding kinetics are measured with buffers starting from 0 M denaturant (native buffer) to increasing denaturant concentrations. The last reaction is measured with a final denaturant concentration just below the equilibrium transition midpoint.

10) The last measurement is the protein baseline of the unfolded protein: the buffer syringe has to be filled with a denaturant buffer of the same concentration as the protein solution, so that protein remains in the unfolded state.

The protein baselines are important as they provide the value of the starting signal of the kinetic trace. In the case of 1:10 mixing ratios, the dead time of the instrument is 3 ms. During this dead time a part of the un/refolding reaction can take place and will not be monitored with the

Protocols

photomultiplier. When the protein baseline signal is known, the amplitude of the full reaction can be determined, and any very fast phase in the dead time can be qualitatively detected.

Further important issues:

- Attachment of the tubings (flow lines) in the KSHU (kinetic syringe holding unit, water bath temperature controlled):

When removing the cell block (metal unit containing the cuvette), the tubings connecting the cuvette with the syringes have to be unscrewed. When the cell block is fit to the instrument again, the tubings have to be attached in such a way, that there is not too much tension on them as the cell block is moved when changing the path length. The best solution is the following: shorter tubing from the protein syringe to the cuvette, longer tubing from the buffer syringe to the cuvette in order to save protein solution.

- Positioning of the flow line entrances into the cuvette

There is no logical explanation, but experiences from other groups and the company Applied Photophysics showed that the best kinetic traces are obtained when both the solutions, the viscous denaturant solution and the less viscous solution enter on the same level into the cuvette. This case is obtained in the 2 mm path position. Less good results are obtained when one of the solutions enter from the top or the bottom of the cuvette, this is the case when the cuvette is turned to the 10 mm path position.

As the stopped-flow instrument is designed to measure fast kinetics, it makes no sense to measure kinetics longer than 50 or 100 seconds. With longer measurement times diffusion effects can disturb the kinetic trace. The cuvette is open on both sides: the entrance from the syringes and the exit that leads to the stop syringe. When the measurement starts, the stop syringe closes one end of the cuvette, but the other end is still open. That explains why diffusion takes place when long measurements are done.

Minimal diffusion effects for the 10 mm path measurements are observed when the viscous solution enters from the bottom of the cuvette. This means, either the tubings connecting cuvette and syringes or the syringe positions themselves have to be exchanged depending on which solutions are used. In unfolding kinetics the buffer syringe contains the more viscous solution, while in refolding kinetics the protein syringe contains the more viscous solution.

For slower kinetics, manual mixing can be performed with the normal CD spectrometer or fluorimeter.

Protocols

- Denaturant solutions in tubings

Don't leave the instrument with denaturant solutions in cuvette and tubings for longer times: the denaturant crystallizes out and plugs the tubings. Very intensive washing is needed to unblock the tubings later on.

- Determination of the dead time of the instrument

The dead time depends on the mixing ratio determined by the syringe volumes. Therefore, one has to measure the dead time for each specific set up. The typical reaction used is the reduction of 2,6-dichlorophenolindophenol (DCIP) by L-ascorbic acid as described in Tonomura et al.[49] Here 1 mol DCIP reacts with 1.1 mol ascorbic acid at pH 2. If ascorbic acid is in excess, a pseudo-first order reaction takes place and the absorbance intensity decreases exponentially. The rate increases proportionally with the ascorbic acid concentration. Equation (12) describes the relation between the observed rate constant k_{obs} and the absorbance signal.

$$Abs_t = Abs_{init} * \exp(-k_{obs} * t) \qquad (12)$$

If the first observation of the absorbance signal takes place at the dead time t_d, then the absorbance at t_d will depend on k_{obs}, therefore the dead time t_d can be extracted from the relation as follows (equation (13)):

$$\ln(Abs_{td}) = \ln(Abs_{init}) - k_{obs} * t_d \qquad (13)$$

This means, you measure at least five reactions at different concentrations of ascorbic acid, you fit the traces to a single exponential equation in order to obtain the observed rate constants k_{obs} (Fig. 14), you plot $\ln(\Delta Abs)$ versus k_{obs} and you perform a linear regression with these data points (Fig. 15). The slope corresponds to the dead time t_d.

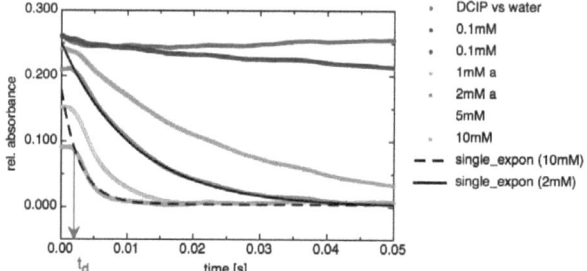

Fig. 14: Kinetic traces of the DCIP – ascorbic acid reaction with five different concentrations of ascorbic acid. The black lines are the best fits to a single exponential equation. The red arrow indicates the dead time t_d.

Fig. 15: Linear regression for the extraction of the dead time t_d.

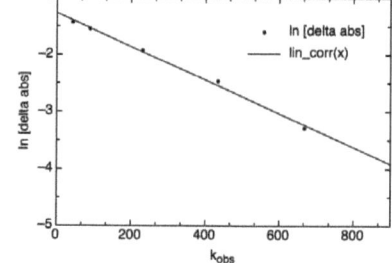

4. Conclusions and Discussion

In this thesis, a series of fully artificial AR proteins have been biophysically characterized. These proteins were previously designed using sequence alignments of natural AR proteins from several protein databases and were termed "full-consensus DARPins".[38] Such a full consensus AR sequence represents an average structure of all known natural AR and therefore serves as a model AR to study the general folding mechanism of AR proteins.

We analyzed a series of six full-consensus DARPins (NI_1C to NI_6C) using the classical cooperative two-state and three-state folding models; furthermore we also developed an alternative (Ising-like) model to describe these proteins. In this model, we assume two different stability values for the repeats: the both capping repeats are less stable than the internal consensus repeats.

Two major insights were obtained: although many AR proteins reveal a two-state behavior in equilibrium folding experiments, all kinetic studies show that the mechanism is more complex. Our generalized AR proteins confirm these results. Second, the stability of the repeat proteins is determined by the unfolding rates. This observation was also made for the full-consensus TRP repeat proteins: with increasing repeat number, the unfolding rates decrease;[45] but in the case of the full-consensus DARPins, the decrease in the unfolding rate was much more significant.

The Ising model could describe our full-consensus DARPin system very well. Using this model, we could extrapolate the stability values for the larger DARPins (NI_4C, NI_5C and NI_6C) as well as a maximal asymptotic denaturation mid-point of even larger DARPins NI_xC ($x \geq 6$). Also, the model allowed predicting the kinetics of NI_1C, NI_2C and NI_3C. As it has been reported for the full-consensus TPR protein series, it is consistent with a more complex folding mechanism than the two-state case. Furthermore, the Ising model provides an energy landscape for each denaturant condition. With this landscape, the most probable folding and unfolding pathways are summarized. The most important observation is the following: at 6 M GdnHCl, the most stable state of NI_3C is the fully denatured one, but also another state almost as stable coexists. This state consists of three folded central repeats and both terminal capping repeats unfolded. Such a detailed interpretation is not possible with the classical cooperative models.

Furthermore, the kinetic unfolding phases can be attributed to different pathways in the energy landscape. In the example of NI_3C, there are three unfolding phases observed in the experiment. The pathway, which contains the states with the lowest energy values (blue pathway in Fig. 6(d)), can be contributed to the slowest unfolding phase. A fast unfolding phase can be explained by another pathway, where a state with higher energy value is crossed (green pathway in Fig. 6(d)).

However, of course, further experimental data are needed to study the intermediate states in more detail. A possible technique is NMR spectroscopy. With this method the proteins can be studied in solution and in diverse solution conditions at equilibrium as well as in a time dependent

Conclusions and perspectives

manner (kinetic measurements). As NMR can give structural information, this is the technique of choice to study folding intermediates. In our case, the complication is the repetitive structure. As the internal repeats of the full-consensus DARPins have all the identical sequence, NMR peaks of these residues cannot be distinguished. A solution of this problem can be the selective labeling of single repeats. However, the construction of a DARPin with selective labeled repeats is technically not trivial and will be a future challenge. If the specific repeat labeling issue will be solved, it would be easy to monitor the H/D exchange of specific repeats. With this tool, equilibrium unfolding curves can be measured using NMR spectroscopy and more detailed informations of the structure of the intermediate states could be obtained.

Another technique used to analyze the intermediate states more in detail, is the stopped-flow double jump experiment (interrupted folding and unfolding assays). This assay allows to examine whether the same intermediate is formed during unfolding and refolding.[25] For the double jump unfolding assay, samples are withdrawn from the unfolding solution at different time intervals after the initiation of unfolding; these samples are transferred to standard refolding conditions, and the amplitudes of the biphasic refolding reactions are determined and the ratios of the amplitudes compared. The amplitudes are proportional to the concentrations of the intermediate species. However, this assay cannot be used, when the both folding steps show similar rates. In this case the amplitudes no longer reflect the concentration of the intermediate molecules. Also, this assay does not give any structural information about the intermediate states.

Apart from the fact that the full consensus design yielded extremely stable proteins when compared to natural AR proteins and to the unselected and selected DARPin library members, another conclusion was that the C-terminal capping repeat is the limiting part for the stability of the whole DARPin. This is an important issue for the design of even more stable DARPins that are useful in biochemical applications, as e. g. DARPins as detection tools.

In this thesis theory and experiment were combined by collaborations with physicists. It was shown that together, theoretical and experimental results open ways to deeper understanding of the 'folding problem'. Both results alone are less meaningful. There is still plenty of space for further experiments.

Conclusions and perspectives

5. Outlook and Perspectives

Several projects are currently ongoing and will be discussed in this paragraph.

The C-cap mutants NI_3C Mut 5, NI_3C Mut 6 and all six NI_1C Mutants are interesting proteins for crystal structure determination. Analysis of the electrostatic interactions in these C-cap mutants could give further insight into the factors that determine the higher stability of these in comparison to other mutants containing fewer point mutations as well as in comparison to the DARPins containing the wild-type C-cap.

Furthermore, NI_3C Mut 5 and NI_3C Mut 6 form dimers to some minor extent (15 %), as seen in static light scattering and gel filtration experiments. Determination of the crystal structure might shed light on the interaction positions that form the protein-protein dimer. Such information could be subsequently be used for further design purpose.

The stable C-cap Mutant 5 can now also be used to increase the stability of other DARPins that are less stable and therefore prone to oligomerize. A first trial is started with an EpCAM-binding DARPin by substituting the C-cap with C-cap mutant 5. The transmembrane glycoprotein EpCAM is an epithelial cell adhesion molecule that shows abundant expression in many solid tumors, but only limited distribution in normal epithelial tissues. EpCAM-specific antibody therapeutics were shown to be promising in preclinical as well as clinical studies.[50] Substituting the wild-type C-cap against the C-cap mutant 5 in EpCAM-specific high-affinity binders might decrease the tendency to form aggregates. The binders with wild-type C-cap will be compared to the binders with the new C-cap mutant 5. This approach might provide DARPin binders well suited as alternative scaffolds for tumor drug delivery.

Another currently ongoing project is the use of NI_6C as a bulky molecular linker. By employing a full-consensus DARPin as linker, inflexible ErbB2-targeted DARPin dimers will be constructed. ErbB2 is another transmembrane protein (a tyrosine-kinase coreceptor) that is abundantly expressed in breast and ovarian tumors.

In order to mimic the structure of antibody arms with simple construction tools, we decided to introduce a bulky protein bridge between two DARPin molecules. For this purpose, we selected the nonspecific consensus DARPin molecules with at least three-internal repeat length (i.e. NI_3C and longer). The two ErbB2-specific binders are connected *via* a very short linker consisting of two residues to the N- and C-termini of the structured part of a nonspecific DARPin. This approach is expected to yield dimeric DARPin molecules where the specificity-conferring moieties are separated by a sufficient distance (comparable to antibodies in the range of 30-100 Å).

Conclusions and perspectives

Nevertheless, sufficient degrees of flexibility and rotation are necessary, as the exact epitope and orientation of DARPins in complex with ErbB2 receptor are not yet known. Instead of a short linker, a longer linker of $(G_4S)_4$ as well as the full-consensus DARPin NI_6C is used.

To investigate whether full-consensus DARPins of varying internal repeat numbers can be used as rigid standard proteins for quantitative distance measurements in single-molecule fluorescence experiments using Förster resonance energy transfer (FRET), FRET transfer efficiency distributions were measured for donor and acceptor dyes attached to the termini of freely diffusing DARPins. These constructs have been previously constructed for the attachment of a donor and an acceptor fluorophore at defined positions within the proteins. In two DARPins containing one and two internal repeats all lysine residues, except for one close to the N-terminus, have been replaced with arginines and an additional C-terminal cysteine has been introduced. In a third DARPin, NI_3C, containing three internal repeats, both a N- and C-terminal cysteine have been introduced.

The project comprised the construction (done by the diploma student J. Kirchholtes), the purification of the DARPins, their conjugation with fluorescent dyes and subsequent single-molecule FRET experiments under native and denaturing conditions as well as the determination of the effect of their modification on their thermodynamic stability by ensemble fluorescence experiments. The single-molecule FRET technique promises new insights into the folding and unfolding mechanism of proteins; however, it remains an open question, whether the full-consensus DAPRins are useful molecules to further fine-tune this method.

In order to further validate the Ising-like folding model of our full-consensus DARPins as well as to explore further possibilities to gain informations about possible folding intermediates, a collaboration with Oliver Zerbe was started.

Folding and formation of hydrogen bonds are intuitively linked. The measurement of exchange kinetics of amide protons therefore presents a convenient method for studying protein folding. In principle, data for H/D exchange could be derived from NMR or MS spectroscopy. ^{15}N labeling of proteins in combination with $^{15}N,^{1}H$ correlation spectroscopy enables us to follow amide proton exchange with residue-resolution. These experiments are sensitive and applicable for larger proteins (i.e. proteins of more than 250 residues). The six full-consensus DARPins will be expressed in ^{15}N medium for uniform labeling of the N-nuclei. The lyophilized two most stable NI_3C C-cap mutants will be dissolved in D_2O for H/D exchange measurements and $^{15}N,^{1}H$ correlation experiments recorded repeatedly to follow the signal decay due to the replacement of amide protons by deuterium. The exchange rates should depend on the stability of the hydrogen bonds, which are related to the overall stability of the folded protein, but also depend on the folding mechanism. If we assume a cooperative model, partial unfolding of the highly stable NI_6C is

Conclusions and perspectives

extremely rare and therefore H/D exchange will be very slow. In contrast, NI_1C is less stable and unfolds at intermediate temperatures, so H/D exchange is expected to take place more rapidly. However, if folding occurs according to the Ising model, where we assume that each repeat unfolds and refolds independently, the length of the protein as well as the stability of each internal repeat is expected to influence the exchange rate only marginally. Nevertheless, the stability of the capping repeats is different, and this difference is expected to yield similar exchange kinetics for the caps in both NI_1C as well as in NI_6C.

It would also be interesting to measure H/D exchange of NI_3C in 3 M GdnHCl in order to reveal details of the equilibrium intermediate states.

If we label the C atoms of NI_3C and the other full-consensus DARPins using ^{13}C medium, sequence-specific assignments might be possible because of the different sequences of the terminal capping repeats, and hence structural information of the capping repeats could be obtained and H,D exchange rates could be assigned to individual sites.

As little information on structures of the folding intermediates of full-consensus DARPins is presently available, a lot of work remains to be done. NMR could turn out to be a valuable method to reveal such information.

Technological aspects, such as the usage of highly stable DARPins as detection tools and alternative scaffolds for drug delivery, will be further explored.

Acknowledgments

My thesis turned out to be a very successful project after a quite long lag phase. That's why I want to thank Andreas Plückthun very much to allow me working on the DARPins (and to push me in the beginning) and for all the stimulating ideas and fruitful discussions about the project and possible collaborations. I learned a lot of new things, i.e. scientifically as well as humanly. Only when you are about to finish such a life period, as it is about to happen with my PhD thesis, you become aware of it and I am very glad about this. I want to thank Ilian Jelesarov and Ben Schuler for all important and helpful discussions we had together about protein folding and science, you were indispensable for my work.

I also want to thank Manca Kenig, she was the next very important person during my PhD, without her I could not have imagined the life in the laboratory, and especially the hard times during my stopped-flow measurements. She turned out to become my "unofficial PhD supervisor" and even more later on, what I call a very good friend.

For the very good collaborations I want to thank Gianluca Interlandi, Giovanni Settanni and Amedeo Caflisch, but then also Armin Hoffmann, Tobias Merz, Peer Mittl, Michaela Kramer and Lewyn Li.

I want to thank Sathya Devi, Laura Meier, Oksana Okhrimenko, Sasa Belic, Sonja Geister, Daniel Nettels, Frank Küster, Frank Hillger that are all members of the neighboring groups working specifically in protein folding. In our group I want to thank Kaspar Binz, Patrik Forrer, Petra Parizek, Daniela Röthlisberger, Michael Stumpp, Martin Kawe, Fabio Parmeggiani, Reto Kolly, Myriam Vincent, Anja Mohr, Zuzanna Langer, Jonas Schäfer, Gautham Varadamsetty, Lutz Kummer, Thomas Huber, Birgit Dreier, Bernhard Schimmele, Béatrice Luginbühl, Alain-Daniel Malebranche, Davide Ferrari, Daniela Bukatz, Alexander Batyuk, Greg Stevens and Alexandre Mooser. I would like to thank Marcel Maier for nice chats in M 96. I also would like to thank my students Christian Winterflood and Antoine Buetti for the nice short time spent together and Janine Kirchholtes for giving me the opportunity to supervise a diploma student for the first time. Then I would like to thank Petra Vogt and Peter Lindner for the organization help, Branka from the Mensa team for her moral support at every lunch break, Maria for each nice smile in the morning coming to my office.

My biggest "thank you" holds to my boyfriend Enrico Guarnera, that supported me in every sense and most of all and at all for the past years; without him I would not have made it...as well as my non-scientific friends Vânia, Benoît, Antje, Gaby, Anna, Annette, Julia, Karine, Berrak, Sinan, Christian, Eva and Clara and all those that I cannot and don't want to mention here.

6. References

1. Marcotte, E. M., Pellegrini, M., Yeates, T. O. & Eisenberg, D. (1999). A census of protein repeats. *J. Mol. Biol.* **293**, 151-60.
2. Kobe, B. & Kajava, A. V. (2000). When protein folding is simplified to protein coiling: the continuum of solenoid protein structures. *Trends Biochem. Sci.* **25**, 509-15.
3. Andrade, M. A., Perez-Iratxeta, C. & Ponting, C. P. (2001). Protein repeats: structures, functions, and evolution. *J. Struct. Biol.* **134**, 117-31.
4. Michaely, P., Tomchick, D. R., Machius, M. & Anderson, R. G. (2002). Crystal structure of a 12 ANK repeat stack from human ankyrinR. *J* **21**, 6387-96.
5. Lux, S. E., John, K. M. & Bennett, V. (1990). Analysis of cDNA for human erythrocyte ankyrin indicates a repeated structure with homology to tissue-differentiation and cell-cycle control proteins. *Nature* **344**, 36-42.
6. Bork, P. (1993). Hundreds of ankyrin-like repeats in functionally diverse proteins: mobile modules that cross phyla horizontally? *Proteins* **17**, 363-74.
7. Mosavi, L. K., Cammett, T. J., Desrosiers, D. C. & Peng, Z. Y. (2004). The ankyrin repeat as molecular architecture for protein recognition. *Protein Sci.* **13**, 1435-48.
8. Radford, S. E. (2000). Protein folding: progress made and promises ahead. *Trends Biochem. Sci.* **25**, 611-8.
9. Anfinsen, C. B., Haber, E., Sela, M. & H., W. F. (1961). The Kinetics of formation of native Ribonuclease during oxidation of the reduced polypeptide chain. *PNAS* **47**.
10. Anfinsen, C. B. (1972). Studies on the principles that govern the folding of protein chains. Nobel Lecture, National Institute of Health.
11. Levinthal, C. (1968). Are there pathways for protein folding? *Journal de Chimie Physique et de physico-chimie biologique* **65**, 44.
12. Dill, K. A. & Chan, H. S. (1997). From Levinthal to pathways to funnels. *Nat Struct Biol* **4**, 10-9.
13. Jelesarov, I. & Bosshard, H. R. (2004). Thermodynamics and Kinetics of Protein Folding, pp. 42. University of Zürich, Biochemistry Department.
14. Capaldi, A. P., Shastry, M. C., Kleanthous, C., Roder, H. & Radford, S. E. (2001). Ultrarapid mixing experiments reveal that Im7 folds via an on-pathway intermediate. *Nat Struct Biol* **8**, 68-72.
15. Jemth, P., Gianni, S., Day, R., Li, B., Johnson, C. M., Daggett, V. & Fersht, A. R. (2004). Demonstration of a low-energy on-pathway intermediate in a fast-folding protein by kinetics, protein engineering, and simulation. *Proc. Natl Acad. Sci. U S A* **101**, 6450-5.
16. Bachmann, A. & Kiefhaber, T. (2001). Apparent two-state tendamistat folding is a sequential process along a defined route. *J. Mol. Biol.* **306**, 375-86.
17. Schmid, F. X. (1983). Mechanism of folding of ribonuclease A. Slow refolding is a sequential reaction via structural intermediates. *Biochemistry* **22**, 4690-6.
18. Buchner, J., Kiefhaber, T. (2005). Protocols - Analytical Solutions of Three-state Protein Folding Models. In *Protein Folding Handbook. Part 1.*, Vol. Volume 1, pp. 402 - 406. WILEY-VCH Verlag GmbH & Co. KGaA, Weinheim.
19. Tang, K. S., Guralnick, B. J., Wang, W. K., Fersht, A. R. & Itzhaki, L. S. (1999). Stability and folding of the tumour suppressor protein p16. *J. Mol. Biol.* **285**, 1869-86.
20. Fersht. (1999). *Structure and Mechanism in protein science: a guide to enzyme catalysis and protein folding*, Freeman, New York.
21. Tang, K. S., Fersht, A. R. & Itzhaki, L. S. (2003). Sequential unfolding of ankyrin repeats in tumor suppressor p16. *Structure (Camb)* **11**, 67-73.

References

22. Interlandi, G., Settanni, G. & Caflisch, A. (2006). Unfolding transition state and intermediates of the tumor suppressor p16INK4a investigated by molecular dynamics simulations. *Proteins* **64**, 178-92.
23. Zhang, B. & Peng, Z. (2000). A minimum folding unit in the ankyrin repeat protein p16(INK4). *J. Mol. Biol.* **299**, 1121-32.
24. Zeeb, M., Rosner, H., Zeslawski, W., Canet, D., Holak, T. A. & Balbach, J. (2002). Protein folding and stability of human CDK inhibitor p19(INK4d). *J. Mol. Biol.* **315**, 447-57.
25. Low, C., Weininger, U., Zeeb, M., Zhang, W., Laue, E. D., Schmid, F. X. & Balbach, J. (2007). Folding Mechanism of an Ankyrin Repeat Protein: Scaffold and Active Site Formation of Human CDK Inhibitor p19(INK4d). *J. Mol. Biol.* **373**, 219-31.
26. Zweifel, M. E., Leahy, D. J., Hughson, F. M. & Barrick, D. (2003). Structure and stability of the ankyrin domain of the Drosophila Notch receptor. *Protein Sci.* **12**, 2622-32.
27. Zweifel, M. E. & Barrick, D. (2001). Studies of the ankyrin repeats of the Drosophila melanogaster Notch receptor. 2. Solution stability and cooperativity of unfolding. *Biochemistry* **40**, 14357-67.
28. Bradley, C. M. & Barrick, D. (2002). Limits of cooperativity in a structurally modular protein: response of the Notch ankyrin domain to analogous alanine substitutions in each repeat. *J. Mol. Biol.* **324**, 373-86.
29. Tripp, K. W. & Barrick, D. (2003). Folding by consensus. *Structure (Camb)* **11**, 486-7.
30. Tripp, K. W. & Barrick, D. (2004). The tolerance of a modular protein to duplication and deletion of internal repeats. *J. Mol. Biol.* **344**, 169-78.
31. Bradley, C. M. & Barrick, D. (2005). Effect of multiple prolyl isomerization reactions on the stability and folding kinetics of the notch ankyrin domain: experiment and theory. *J. Mol. Biol.* **352**, 253-65.
32. Bradley, C. M. & Barrick, D. (2006). The notch ankyrin domain folds via a discrete, centralized pathway. *Structure* **14**, 1303-12.
33. Tripp, K. W. & Barrick, D. (2007). Enhancing the Stability and Folding Rate of a Repeat Protein through the Addition of Consensus Repeats. *J. Mol. Biol.* **365**, 1187-200.
34. Street, T. O., Bradley, C. M. & Barrick, D. (2007). Predicting coupling limits from an experimentally determined energy landscape. *Proc. Natl Acad. Sci. U S A* **104**, 4907-12.
35. Lubman, O. Y., Kopan, R., Waksman, G. & Korolev, S. (2005). The crystal structure of a partial mouse Notch-1 ankyrin domain: repeats 4 through 7 preserve an ankyrin fold. *Protein Sci.* **14**, 1274-81.
36. Mello, C. C. & Barrick, D. (2004). An experimentally determined protein folding energy landscape. *Proc. Natl Acad. Sci. U S A* **101**, 14102-7.
37. Mello, C. C., Bradley, C. M., Tripp, K. W. & Barrick, D. (2005). Experimental characterization of the folding kinetics of the notch ankyrin domain. *J. Mol. Biol.* **352**, 266-81.
38. Binz, H. K., Stumpp, M. T., Forrer, P., Amstutz, P. & Plückthun, A. (2003). Designing repeat proteins: well-expressed, soluble and stable proteins from combinatorial libraries of consensus ankyrin repeat proteins. *J. Mol. Biol.* **332**, 489-503.
39. Stumpp, M. T., Forrer, P., Binz, H. K. & Pluckthun, A. (2003). Designing repeat proteins: modular leucine-rich repeat protein libraries based on the mammalian ribonuclease inhibitor family. *J. Mol. Biol.* **332**, 471-87.
40. Parmeggiani, F., Pellarin, R., Larsen, A. P., Varadamsetty, G., Stumpp, M. T., Zerbe, O., Caflisch, A. & Pluckthun, A. (2008). Designed armadillo repeat proteins as general peptide-binding scaffolds: consensus design and computational optimization of the hydrophobic core. *J. Mol. Biol.* **376**, 1282-304.
41. Forrer, P., Stumpp, M. T., Binz, H. K. & Pluckthun, A. (2003). A novel strategy to design binding molecules harnessing the modular nature of repeat proteins. *FEBS Lett* **539**, 2-6.

References

42. Forrer, P., Binz, H. K., Stumpp, M. T. & Pluckthun, A. (2004). Consensus design of repeat proteins. *Chembiochem* **5**, 183-9.
43. Main, E. R., Xiong, Y., Cocco, M. J., D'Andrea, L. & Regan, L. (2003). Design of stable alpha-helical arrays from an idealized TPR motif. *Structure (Camb)* **11**, 497-508.
44. Main, E. R., Jackson, S. E. & Regan, L. (2003). The folding and design of repeat proteins: reaching a consensus. *Curr Opin Struct Biol* **13**, 482-9.
45. Main, E. R., Lowe, A. R., Mochrie, S. G., Jackson, S. E. & Regan, L. (2005). A recurring theme in protein engineering: the design, stability and folding of repeat proteins. *Curr Opin Struct Biol* **15**, 464-71.
46. Main, E. R., Stott, K., Jackson, S. E. & Regan, L. (2005). Local and long-range stability in tandemly arrayed tetratricopeptide repeats. *Proc. Natl Acad. Sci. U S A* **102**, 5721-6.
47. Kajander, T., Cortajarena, A. L., Main, E. R., Mochrie, S. G. & Regan, L. (2005). A new folding paradigm for repeat proteins. *J. Am. Chem. Soc.* **127**, 10188-90.
48. Eftink, M. R. (1994). The use of fluorescence methods to monitor unfolding transitions in proteins. *Biophys J* **66**, 482-501.
49. Monsellier, E. & Bedouelle, H. (2005). Quantitative measurement of protein stability from unfolding equilibria monitored with the fluorescence maximum wavelength. *Protein Eng Des Sel* **18**, 445-56.
50. Tonomura, B., Nakatani, H., Ohnishi, M., Yamaguchi-Ito, J. & Hiromi, K. (1978). Test reactions for a stopped-flow apparatus. Reduction of 2,6-dichlorophenolindophenol and potassium ferricyanide by L-ascorbic acid. *Anal Biochem* **84**, 370-83.
51. Hussain, S., Pluckthun, A., Allen, T. M. & Zangemeister-Wittke, U. (2007). Antitumor activity of an epithelial cell adhesion molecule targeted nanovesicular drug delivery system. *Mol Cancer Ther* **6**, 3019-27.

7. Appendix

7.1 Abbreviations

aa	amino acid
AFM	atomic force microscopy
ANK	ankyrin
AR	ankyrin repeat
CD	circular dichroism
Chevron plot	graph where the logarithm of the folding and unfolding rate is plotted versus the denaturant concentration
DARPins	designed AR proteins
Dm	midpoint of denaturant concentration in chemical unfolding
ΔG	difference in free energy between two states
ΔH	difference in free enthalpy between two states
GA	guanine-adenine
GABPβ1	mouse guanine-adenine-binding protein β1 subunit
GdnHCl	guanidinium hydrocloride
HB	hydrogen bond
IPTG	Isopropyl-β-thiogalactopyranoside
LRR	Leucine-rich repeat
NMR	Nuclear magnetic resonance
MALS	Multi-Angle-Light Scattering
MD	molecular dynamics
PCR	Polymerase Chain reaction
PDB	Protein Data Bank
PFAM	Protein families database of alignments and hidden Markov models
RMSD	root mean square deviation
SC	surface complementarity
SDS	Sodium dodecyl sulphate
SMART	Simple modular architecture research tool
Tm	midpoint temperature in thermal unfolding
TPR	tetratricopeptide repeat
TPRA1	mechanosensitive transduction channel

Appendix

7.2 List of Plasmids

Plasmids used during the thesis:

pPANK (HKB)
pPRO (HKB)
pEWT (HKB)
pWTC (HKB)
pSW_N1C
pSW_N2C
pSW_N3C
pSW_N4C
pSW_N5C
pSW_N6C
pSW_I3
pSW_I4
pSW_NI3
pSW_NI4
pSW_I3C
pSW_I4C
pSW_N1CMu1
pSW_N1CMu2
pSW_N1CMu3
pSW_N1CMu4
pSW_N1CMu5
pSW_N1CMu6
pSW_N3CMu1
pSW_N3CMu2
pSW_N3CMu3
pSW_N3CMu4
pSW_N3CMu5
pSW_N3CMu6
pJKN1CKR
pJKN2CKR
pJKmuA

Appendix: Equations

7.3 Equations used for fitting of cooperative folding models

For fitting the experimental data the software ProFit 6.0.6 was used. The equations are represented in a Pascal-like function definition.

7.3.1 Equilibrium two-state fit
7.3.1.1 Chemical unfolding, fitting parameters m, Dm

```
function equil_2state_chem;

description
'y = ((m1+m2*x)+(m3+m4*x)*exp(m5*(x-m6)/R*T))/(1+exp(m5*(x-m6)/R*T));)',
'2-state equil_chem';

defaults
        a[1]:=0,active,'yf (signal folded)';
        a[2]:=0,active,'mf (slope of pretransition)';
        a[3]:=0,active,'yu (signal unfolded)';
        a[4]:=0,active,'mu (slope of posttransition)';
        a[5]:=0,active,'m (slope of transition)';
        a[6]:=0,active,'Dm (transition midpoint)';
        a[7]:=0,inactive,'R (gas constant)';
        a[8]:=0,inactive,'T (temp)';

begin
        y := ((a[1]+a[2]*x)+(a[3]+a[4]*x)*exp(a[5]*(x-a[6])/(a[7]*a[8])))/(1+exp(a[5]*(x-a[6])/(a[7]*a[8])))
end;
```

7.3.1.2 Chemical unfolding, fitting parameters m, ΔG^0

```
function equ_2state_chem_deltaG;

description
'equil 2-state chem';

defaults
        a[1]:=-13000,active,'An';
        a[2]:=120,active,'mn';
        a[3]:=-1270,active,'Bu';
        a[4]:=40,active,'mu';
        a[5]:=9000,active,'dG0';
        a[6]:=2500,active,'m';
        a[7]:=293.15, inactive, 'T';

var

denat, dG, K, fn, fu, Sn, Su;

begin
        denat := x;
        Sn:= a[1] + a[2]*denat;
        Su:= a[3] + a[4]*denat;
        dG := a[5] - a[6]*denat;
        K := exp(-dG/(1.9872*a[7]));
```

Appendix: Equations

```
        fu:= K/(1 + K);
        fn := 1 - fu;
        y  := Sn*fn + Su*fu;
end;
```

7.3.1.3 Thermal unfolding, fitting parameters m, ΔH^0

```
function equil_2state_melting;

description
'2-state equil_melting';

defaults
        a[1]:=0,active,'yu (signal unfolded)';
        a[2]:=0,active,'mu (slope of posttransition)';
        a[3]:=0,active,'yf (signal folded)';
        a[4]:=0,active,'mf (slope of pretransition)';
        a[5]:=0,active,'dH (delta H)';
        a[6]:=0,active,'Tm (melting point, K)';
        a[7]:=1.9872,inactive,'R (gas constant, cal/K*mol)';

var

CDN, CDU, Ku, fu, fn;

begin
CDN := a[2]*x+a[1];
CDU := a[4]*x+a[3];
Ku := exp((a[5]/a[7])*(1/a[6]-1/x));
fu := Ku/(1+Ku);
fn := (1-fu);
y  := fn*CDN+fu*CDU;

end;
```

7.3.2 Equilibrium three-state fit

7.3.2.1 Chemical unfolding, fitting parameters m_1, ΔG_1^0, m_2, ΔG_2^0

```
function equ_3state_chem_deltaG;

description
'equil 3-state chem';

defaults
        a[1]:=600,active,'An';
        a[2]:=-3,active,'mn';
        a[3]:=0.001,active,'Bi';
        a[4]:=1.2,active,'mi';
        a[5]:=7000,active,'Cu';
        a[6]:=-3,active,'mu';
        a[7]:=0.001,active,'dG01';
        a[8]:=0.7,active,'dG02';
        a[9]:=0.5, active,'m1';
        a[10]:=0.5, active,'m2';
        a[11]:=293.15, inactive, 'T';

var
```

Appendix: Equations

```
denat, dG1, dG2, K1, K2, fn, fi, fu, Sn, Si, Su;

begin
        denat := x;
        Sn:= a[1] + a[2]*denat;
        Si:= a[3] + a[4]*denat;
        Su:= a[5] + a[6]*denat;
        dG1 := a[7] - a[9]*denat;
        dG2 := a[8] - a[10]*denat;
        K1 := exp(-dG1/(1.9872*a[11]));
        K2 := exp(-dG2/(1.9872*a[11]));
        fn := 1/(1 + K1 + K1*K2);
        fi := K1/(1 + K1 + K1*K2);
        fu:= K1*K2/(1 + K1 + K1*K2);

        y := Sn*fn + Si*fi + Su*fu;
end;
```

7.3.2.2 Chemical unfolding, fitting parameters m_1, K_1^0, m_2, K_2^0

```
function Equil_3_state;
defaults

a[1] := 90, active, 'Su';
a[2] := 3, active, 'u';
a[3] := 15, active, 'K°1';
a[4] := 15, active, 'K°2';
a[6] := 6.5, active, 'm2';
a[5] := 6.5, active, 'm1';
a[7] := 293.15, inactive, 'T';
a[8] := 90, active, 'Si';
a[9] := 500, active, 'Sn';
a[10] := -30, active, 'n';

begin

y :=
((a[1]+a[2]*x)*a[3]*a[4]*exp((a[5]+a[6])*x/1.987/a[7])+a[8]*a[3]*exp(a[5]*x/1.987/a[7])+a[9]+a[10]*x
)/(1+a[3]*exp(a[5]*x/1.987/a[7])+a[3]*a[4]*exp((a[5]+a[6])*x/1.987/a[7]));end;

end;
```

7.3.3 Kinetics: Two-state chevron plot

```
function chevron_2state;

description
'y = ln((kref*exp(mref*x)) + kunf*exp(munf*x))',
'2-state chevron';

defaults
        a[1]:=0,active,'kref (refolding rate)';
        a[2]:=0,active,'mref (slope of refolding limb)';
        a[3]:=0,active,'kunf (unfolding rate)';
        a[4]:=0,active,'munf (slope of unfolding limb)';
```

Appendix: Equations

```
begin
        y := ln((a[1]*exp(a[2]*x)) + a[3]*exp(a[4]*x))
end;
```

7.3.4 Kinetics: Three-state chevron plot
7.3.4.1 on-pathway intermediate

```
function kin_3state_on_pathway;

description
'3-state on_pathway';

defaults
        a[1]:=600,inactive,'k12';
        a[2]:=-3,inactive,'m12';
        a[3]:=0.001,inactive,'k21';
        a[4]:=1.2,inactive,'m21';
        a[5]:=7000,inactive,'k23';
        a[6]:=-3,inactive,'m23';
        a[7]:=0.001,inactive,'k32';
        a[8]:=0.7,inactive,'m32';

var

denat, kUI, kIU, kIN, kNI, B, C, amp1, amp2;

begin

if (x<10) and (x>=0) then begin
        denat := x;
        kUI := a[1]*exp(a[2]*denat);
        kIU := a[3]*exp(a[4]*denat);
        kIN := a[5]*exp(a[6]*denat);
        kNI := a[7]*exp(a[8]*denat);
        B := -(kUI + kIU + kIN + kNI);
        C := kUI*kIN + kUI*kNI + kIU*kNI;
        amp1 := (-B-sqrt(B^2-4*C))/2;
        y := ln(amp1);
end

else if (x<20) and (x>=10) then begin
        denat := x-10;
        kUI := a[1]*exp(a[2]*denat);
        kIU := a[3]*exp(a[4]*denat);
        kIN := a[5]*exp(a[6]*denat);
        kNI := a[7]*exp(a[8]*denat);
        B := -(kUI + kIU + kIN + kNI);
        C := kUI*kIN + kUI*kNI + kIU*kNI;
        amp2 := (-B+sqrt(B^2-4*C))/2;

        y := ln(amp2);
end

end;
```

7.3.4.2 off-pathway intermediate

Appendix: Equations

```
function kin_3state_off_pathway;

description
'3-state off_pathway';

defaults
      a[1]:=600,active,'k12';
      a[2]:=-3,active,'m12';
      a[3]:=0.001,active,'k21';
      a[4]:=1.2,active,'m21';
      a[5]:=7000,active,'k23';
      a[6]:=-3,active,'m23';
      a[7]:=0.001,active,'k32';
      a[8]:=0.7,active,'m32';

var

denat, kIU, kUI, kUN, kNU, B, C, amp1, amp2;

begin

if (x<10) and (x>=0) then begin
      denat := x;
      kIU := a[1]*exp(a[2]*denat);
      kUI := a[3]*exp(a[4]*denat);
      kUN := a[5]*exp(a[6]*denat);
      kNU := a[7]*exp(a[8]*denat);
      B := -(kUI + kIU + kUN + kNU);
      C := kIU*kUN + kIU*kNU + kUI*kNU;
      amp1 := (-B-sqrt(B^2-4*C))/2;
      y := ln(amp1);
end

else if (x<20)  and (x>=10) then begin
      denat := x-10;
      kIU := a[1]*exp(a[2]*denat);
      kUI := a[3]*exp(a[4]*denat);
      kUN := a[5]*exp(a[6]*denat);
      kNU := a[7]*exp(a[8]*denat);
      B := -(kUI + kIU + kUN + kNU);
      C := kIU*kUN + kIU*kNU + kUI*kNU;
      amp2 := (-B+sqrt(B^2-4*C))/2;

      y := ln(amp2);
end

end;
```

7.3.4.3 triangular pathway intermediate

```
function kin_3state_triangular;

description
'3-state triangular';

defaults
      a[1]:=600,active,'k12';
      a[2]:=-3,active,'m12';
      a[3]:=0.001,active,'k21';
      a[4]:=1.2,active,'m21';
      a[5]:=7000,active,'k23';
```

Appendix: Equations

```
        a[6]:=-3,active,'m23';
        a[7]:=0.001,active,'k32';
        a[8]:=0.7,active,'m32';
        a[9]:=7000,active,'k13';
        a[10]:=-3,active,'m13';
        a[11]:=0.001,active,'k31';
        a[12]:=0.7,active,'m31';

var

denat, kIU, kUI, kUN, kNU, kIN, kNI, B, g1, g2, g3, C, amp1, amp2;


begin

if (x<10) and (x>=0) then begin
        denat := x;
        kUI := a[1]*exp(a[2]*denat);
        kIU := a[3]*exp(a[4]*denat);
        kIN := a[5]*exp(a[6]*denat);
        kNI := a[7]*exp(a[8]*denat);
        kUN := a[9]*exp(a[10]*denat);
        kNU := a[11]*exp(a[12]*denat);

        B := -(kUI + kIU + kUN + kNU + kIN + kNI);
        g1 := kNI*kIU + kNU*kIU + kIN*kNU;
        g2 := kNU*kUI + kUI*kNI + kUN*kNI;
        g3 := kUI*kIN + kIU*kUN + kUN*kIN;

        C := g1 + g2 +g3;
        amp1 := (-B-sqrt(B^2-4*C))/2;
        y := ln(amp1);
end
else if (x<20)  and (x>=10) then begin
        denat := x-10;
        kUI := a[1]*exp(a[2]*denat);
        kIU := a[3]*exp(a[4]*denat);
        kIN := a[5]*exp(a[6]*denat);
        kNI := a[7]*exp(a[8]*denat);
        kUN := a[9]*exp(a[10]*denat);
        kNU := a[11]*exp(a[12]*denat);

        B := -(kUI + kIU + kUN + kNU + kIN + kNI);
        g1 := kNI*kIU + kNU*kIU + kIN*kNU;
        g2 := kNU*kUI + kUI*kNI + kUN*kNI;
        g3 := kUI*kIN + kIU*kUN + kUN*kIN;

        C := g1 + g2 +g3;
        amp2 := (-B+sqrt(B^2-4*C))/2;

        y := ln(amp2);
end

end;
```

Die VDM Verlagsservicegesellschaft sucht für wissenschaftliche Verlage abgeschlossene und herausragende

Dissertationen, Habilitationen, Diplomarbeiten, Master Theses, Magisterarbeiten usw.

für die kostenlose Publikation als Fachbuch.

Sie verfügen über eine Arbeit, die hohen inhaltlichen und formalen Ansprüchen genügt, und haben Interesse an einer honorarvergüteten Publikation?

Dann senden Sie bitte erste Informationen über sich und Ihre Arbeit per Email an *info@vdm-vsg.de*.

Sie erhalten kurzfristig unser Feedback!

VDM Verlagsservicegesellschaft mbH
Dudweiler Landstr. 99
D - 66123 Saarbrücken
Telefon +49 681 3720 174
Fax   +49 681 3720 1749
www.vdm-vsg.de

Die VDM Verlagsservicegesellschaft mbH vertritt

Printed by Books on Demand GmbH, Norderstedt / Germany